The Causal Mindset Handbook

A practical guide to the science of cause and effect for evidence-based decision making

Dr. Quentin Gallea

<packt>

The Causal Mindset Handbook

Portfolio Director: Sunith Shetty
Relationship Lead: Nilesh Kowadkar
Project Manager: Shashank Desai
Content Engineer: Tiksha Abhimanyu Lad
Technical Editor: Seemanjay Ameriya
Copy Editor: Safis Editing
Indexer: Manju Arasan
Proofreader: Tiksha Abhimanyu Lad
Production Designer: Vijay Kamble
Growth Lead: Merlyn M Shelley

First published: February 2026
Production reference: 1050226

Published by Packt Publishing Ltd.
Grosvenor House
11 St Paul's Square
Birmingham
B3 1RB, UK.

ISBN 978-1-80611-785-7
www.packtpub.com

To Cécile, Mia and Nell, who causally contribute profoundly to making life beautiful.

– Dr. Quentin Gallea

Foreword

We live in an era defined by prediction.

Artificial intelligence systems forecast consumer behavior, detect fraud, generate content, and design molecules with increasing precision. Inside firms, algorithms now make millions of micro-decisions every second, from pricing to logistics to customer recommendations. Prediction, once scarce and costly, has become abundant and cheap.

Yet despite this abundance, decision quality has not improved in parallel. Managers and policymakers are surrounded by data, models, and dashboards, but they still struggle to answer the most important question: why. The difference between understanding patterns and understanding causes remains profound. Prediction tells us what is likely to happen. Causality tells us what will happen if we act.

Quentin Gallea's *The Causal Mindset Handbook* arrives at exactly this juncture. It provides a disciplined, accessible, and rigorous framework for causal reasoning in an age dominated by correlations. The book demystifies the science of causal inference and shows how to apply it to real-world problems in business, policy, and everyday life. It is not a technical treatise. It is a guide to better thinking.

The Age of Abundant Expertise

In my book Competing in the Age of AI (Harvard Business Review Press, 2020, with Marco Iansiti), we described how firms build "AI factories" that transform data into predictions, predictions into actions, and actions into learning. These architectures operate at scale and power the modern digital enterprise. But the hidden assumption behind this model is that correlation is good enough. When algorithms start allocating credit, diagnosing patients, or guiding hiring, that assumption becomes risky. Predictive models can only optimize within the current system. Causal reasoning allows us to redesign it.

The same logic applies to strategy. In the Harvard Business Review article “Strategy in an Era of Abundant Expertise” (March–April 2025), co-authored with Bobby Yerramilli-Rao, John Corwin, and Yang Li, we argued that AI is collapsing the cost of access to knowledge. Expertise is no longer scarce. The scarce resource now is the ability to reason about cause and effect in an environment flooded with information. Gallea’s book is about mastering that capability. It teaches how to move from pattern recognition to causal understanding, from reaction to design.

My Causal Journey

My own research journey has been shaped by a desire to understand cause and effect directly rather than indirectly. Trained in management, sociology, and economics, I often felt that traditional empirical work in the social sciences relied too heavily on retrospective inference. I was not content being a fast historian—using econometrics to reconstruct what might have happened after the fact. I wanted to observe innovation and creativity as they unfolded and to test, in real time, how different interventions changed behavior and performance. That impulse led me toward field experimentation as both a method and a mindset.

Over the past two decades, that commitment has defined my research and the institutions I have built. The NASA Tournament Lab was an early proof of concept: could open innovation challenges mobilize distributed expertise to solve frontier problems in aeronautics and space science? The results demonstrated how crowd-based problem solving could outperform traditional hierarchical R&D models. The work expanded into the Crowd Innovation Lab at Harvard, which tested digital platforms for scientific and technical breakthroughs, and later into the **Laboratory for Innovation Science at Harvard (LISH)**, which scaled these methods across industries. Each experiment—from crowdsourcing medical imaging solutions to designing hybrid human–AI decision systems—pushed closer to identifying the causal levers of creativity, coordination, and expertise. These experiences ultimately shaped the ethos of the **Digital, Data, and Design** (D^3) Institute at Harvard, where our mission is to study and accelerate the causal transformation of organizations in the AI era.

From Predictive AI to Causal Intelligence

Predictive AI helps us see the world. Causal intelligence helps us change it.

Most questions that matter are causal. What drives climate change? Does an AI assistant improve or degrade team performance? Will a marketing campaign alter customer behavior or merely coincide with it? These are counterfactual questions that require causal thinking.

Gallea's five-step Causal Mindset Framework provides a practical structure for reasoning about these questions

1. Identify the counterfactual: What is being compared?
2. Search for "something else." What hidden factors might explain the result?
3. Determine the direction of bias. Are effects amplified or attenuated?
4. Refine and iterate. Seek better comparisons and stronger evidence.
5. Test extrapolation. Do results generalize across time and context?

This disciplined process transforms how decisions are framed and evaluated. It replaces intuition with structured curiosity. The most advanced AI organizations now need this kind of reflection built into their workflows. Just as we design feedback loops for algorithms, we must design counterfactual loops for human judgment.

The Causal Leader

The leaders who will succeed in the AI economy are those who pair technological fluency with causal literacy. They frame decisions as experiments, demand clear counterfactuals, invest in interpretable metrics, and accept uncertainty as a source of learning. In our field experiments at Boston Consulting Group and Procter & Gamble, we have seen how this mindset changes organizations. AI increases performance when human judgment is guided by structured experimentation, not when it replaces it.

Gallea's framework equips leaders to apply this discipline across domains. It makes causal reasoning a daily managerial practice rather than a statistical specialty. It creates leaders who understand that good strategy is built not on stories or correlations, but on tested cause-and-effect relationships.

The Architecture of the Frontier Firm

At Harvard Business School we study Frontier Firms that operate at the edge of technological and managerial innovation. These firms integrate technology, data, and organization into a continuous learning system. Gallea's work complements this perspective. His Causal Mindset provides the cognitive architecture that matches the digital one. Where the AI Factory automates prediction, the Causal Mindset automates inquiry. Together they define the operating model of the future firm: one that learns intentionally through structured experimentation and causal feedback.

This synthesis will define the next decade of management research and practice. The most advanced organizations will not just deploy AI systems that predict. They will build causal systems that explain, test, and learn. Their advantage will come not from more data, but from better questions.

The Future Belongs to Causal Thinkers

Every industrial revolution automates something humans once did intuitively. The first automated physical labor. The digital revolution automated information processing. The AI revolution is automating prediction. The next frontier will automate reasoning about causation.

When prediction becomes cheap, causality becomes valuable. When data are abundant, understanding which patterns to trust becomes decisive. The firms and leaders that master causal reasoning will shape the next era of intelligent organizations.

This book is a roadmap for that journey. It helps us see beyond the illusion of correlation, to understand the mechanisms that actually drive outcomes. It reminds us that leadership in the AI age is not about having more models or dashboards. It is about building the capacity to learn, test, and act on what truly causes change.

Prediction allows us to see the world as it is.

Causality allows us to change it.

Karim Lakhani

– Cambridge, MA, Fall 2025, Harvard Business School | Digital Data Design Institute at Harvard | Laboratory for Innovation Science at Harvard

Contributors

About the author

With deep expertise and a passion for causality, **Dr. Quentin Gallea** has published research in top journals, addressing timely issues like the effects of COVID lockdowns or weapon imports on conflict. He is known for his unique ability to make complex topics accessible without compromising rigor. His practical approach has empowered over 15,000 people from C-suites to experienced researchers to apply causal thinking and causal inference in their respective fields. Today, Quentin is self-employed and provides advisory, training, public speaking, and advisory services across the world with a particular focus on measuring the impact of AI. In addition, he is a Senior Advisor at Enlighten Advisory, offering his expertise to support strategic decisions.

Acknowledgment

Countless people contributed to this book; some wrote seminal work, others debated causality with me for numerous hours, or supported my tireless quest to master causal inference and share it with the world. On the other hand, many encouraged me to write this book without realizing it by making public, heroic, misleading causal claims. While I will probably fail to cite everyone, there are still numerous people that I definitely want to mention here.

I would like to thank the economists who inspired my career early on: Esther Duflo and Abhijit Banerjee, whose work and book Poor Economics made me realize, as an undergraduate, the importance of causality in answering timely questions and contributing to society; Josh Angrist for his incredible capacity to communicate and teach econometrics (see the American Economic Association videos); Scott Cunningham for writing one of the most beautiful books on causal inference, The Mixtape, and sharing his expertise online.

Many people directly gave me feedback and helped me with my writing, starting with Matheus Facure, who gave me incredibly valuable feedback using his technical and applied industry knowledge as well as sharing his knowledge in the best industry-oriented books on causal inference; Dr. Ryan Batten for his feedback on directed graphs; Robson Tigre and Justin Bélair for the comments on the randomized experiments chapter; and Walker Hanlon for his double check on the analysis of his paper on fog and pollution in London. In addition to the technical feedback, industry leaders and other technical experts went through an early and messy draft: Rafael Lalive, Gilles Carbonier, Julia Binder, Patrick Dunne, Alexia Michiels, and Konstantinos Apostolatos.

Karim Lakhani enriched the book with his wide and deep expertise from academia and industry. I am grateful for how his foreword provides a vivid illustration of the applicability and necessity of this book.

A special thanks to Charles Ayoubi, who always supported my project, debated causality day and night, challenged my ideas, and pushed me to dig deeper. Of course, I can't forget to also thank Boris Thurm, who shared his expertise in environmental science and impact with humor and rigor. In addition, I want to deeply thank friends who believed in and encouraged me: Sean Seale, Camil Schilliger, and Eleonore Poli.

Without a doubt, I cannot thank enough my thousands of students, in particular those from HEC Lausanne, from whom I learned so much while teaching; as well as my classmates and professors during my master's and Ph.D. in economics.

My editor and team from Packt did an amazing job, and our joint enthusiasm for the project, combined with their valuable feedback, greatly contributed to bringing the book to where it is today: Tiksha Abhimanyu Lad, Nilesh Kowadkar, and Shashank Desai.

But of course, my deepest thanks go directly to my family: my daughters, who are my entire world and made me realize that causal inference can't explain everything and, in particular, that there was something bigger than my intellectual obsession; my wife, who supported my countless projects, crazy ideas, and ups and downs; my older brother, who read encyclopedias to me when we were children and always challenged intellectual shortcuts; and my parents, who allowed me to follow my passion, go to university after an apprenticeship, and be curious.

About the reviewer

Matheus Facure Alves is a Brazilian economist-turned data scientist and Staff Data Scientist at Nubank. He is the author of the open-source classic *Causal Inference for the Brave and True and Causal Inference in Python* (O'Reilly), both translated into multiple languages. He works at the intersection of econometrics and machine learning, with a focus on fintech. His projects span credit policy, A/B testing, pricing, and marketing, emphasizing heterogeneous treatment effects, uplift modeling, and personalization. A code-first educator, he shares reproducible notebooks and teaches workshops. He lives in Campinas, São Paulo, with his family.

Table of Contents

Preface

I am often asked why causality is my passion. Why have I dedicated so much time to working on this and educating people on multiple continents? To me, it is so obvious that I often forget that it might sound surprising. Most of our important questions are about cause and effect: What is the effect of this vaccine? What is the impact of pollution on mortality? What is the cause of this conflict? What is the impact of mobile phones and social media on mental health?

As most of our questions rely on causality, most of our decisions do as well. We decide to implement a policy because we expect an impact (causal effect) on poverty, pollution, or health. We make business decisions expecting positive effects on growth.

Hence, causality is arguably one of the most valuable concepts anyone could learn today to make better decisions and prevent manipulation. However, there is a catch. The scientific way to study causality is often made inaccessible to a wide audience as it requires complex math and statistics. As Joshua Angrist, Nobel Memorial Prize in Economics winner, said:

> *"The traditional econometrics [the study of economic phenomena with statistics, where assessing causality is central] course is relatively abstract. It sort of looks like a course in Statistical Theory and uses a lot of math, maybe proving theorems [...] So, I think the principal barrier to widespread improvement of econometric methods in public life, say outside of Academia, is the lack of good instruction. A college degree is meant to be a lifetime investment for everyone."*

This book aims to address this issue by demystifying the science of causality, making it both accessible and practical for a broad audience.

Who this book is for

These tools will help business executives make better strategic decisions, data scientists work more effectively with data, news reporters and consumers separate facts from fiction, and scientific researchers strengthen their research.

This book can benefit anyone curious to go beyond correlations, understand the "why," and acquire a competitive advantage in the job market with new, valuable skills. *The Causal Mindset* offers an accessible, non-technical path to master causal thinking, improve data literacy, and confidently make better decisions. No math or stats background is necessary, just a curiosity about data.

What this book covers

At the end of my Ph.D. defense, the internal expert asked me about the most important thing I learned during my Ph.D. What do you think I responded with? An equation, a model, some mathematical notions? Of course not! One of the biggest takeaways was that I had a glimpse at the complexity of the world. I became more conservative and humbler concerning my knowledge, but it also allowed me to accept my own limitations and embrace uncertainty. I was somewhat witnessing what we call the Dunning-Kruger effect.

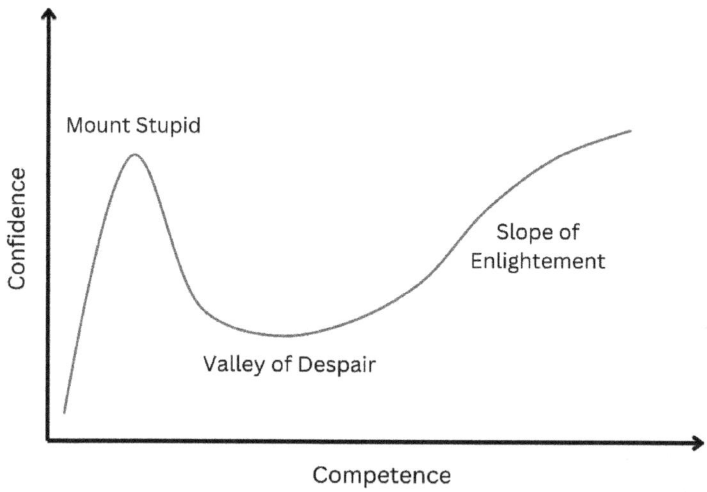

Figure 1: Representation of a theoretical relationship between knowledge on a topic and confidence, called the Dunning-Krueger effect

Let me tell you more about the Dunning-Krueger effect before walking you through the content of this book, as you might experience it as well while reading. Similar to my Ph.D. experience, throughout this book, you will follow the ups and downs of this curve. The Dunning-Krueger effect is a cognitive bias illustrating the fact that people with little knowledge about a topic tend to overestimate their expertise. I am sure that you have had a discussion with someone who was very vocal while having little clue about a certain topic. *Figure 1* illustrates this mechanism by plotting the relationship between confidence in a topic and actual knowledge. First, people tend to quickly get confident about their knowledge when they discover a new subject, reaching what is known as *Mount Stupid*. However, the more they learn, the more they understand the complexity and see how little they know, pushing them down the *Valley of Despair*. Then, slowly, by digging deeper and deeper into the same topic, they follow the *Slope of Enlightenment* to eventually become an expert on this matter. Being an expert does not mean that you know absolutely everything. Sure, part of the story is that you have spent years learning and practicing and you have deep knowledge on one topic. But equally important as the knowledge acquired is that you know its limitations and how to deal with them. Through this book, we will embark on a journey. This journey, represented in the following figure, will reflect what I experienced during my Ph.D., namely, the Dunning-Kruger effect.

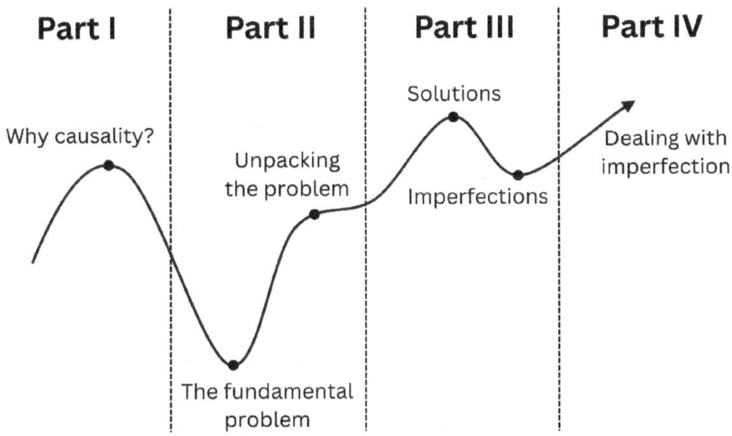

Figure 2: Overview of the book

Part I, composed of *Chapters 1* and *2*, will provide the foundation, allowing us to understand what causality is, its importance, and how it differs from what we often do with AI.

- *Chapter 1, Understanding Causality and Its Importance*, shows what causality is, how to graphically represent it, and why understanding it helps to make better decisions and prevent being subject to misinformation or manipulation.
- *Chapter 2, Causation versus Prediction*, explores the difference between AI made essentially for prediction and causal analysis (including causal AI). We will see what the differences are, when to use which, and how to use them together. Understanding the difference between predictive and causal inference isn't just a technical detail; it's a strategic necessity. Grasping these concepts helps with understanding the limitations of AI, knowing which tool to use and how to allocate resources more effectively, and, ultimately, attaining better outcomes.

At first, we will reach a peak of enthusiasm, discovering just how powerful causal reasoning can be. However, there is a catch: causality is actually incredibly hard to prove. *Part II* of the book, including *Chapters 3* and *4*, will reveal the main challenges.

- *Chapter 3, Why Is It So Hard To Prove Causality?*, dives into the depths of the fundamental problem of causal inference. We will see that the only way to perfectly measure a causal effect is something impossible: having access to parallel worlds. Then, we will start to climb the slope progressively.
- *Chapter 4, Beyond Correlation: The Main Culprits*, identifies the main problems that prevent measuring a causal effect. By identifying those effects, we will equip ourselves with tools that allow us to question causal claims.

Once we've explored what prevents us from making a causal claim, we will turn to *Part III*: the solutions. We will keep climbing the slope of enlightenment and reach a new peak where we will discover different solutions to tackle the problems and eventually measure a causal effect precisely.

- *Chapter 5, The Causal Mindset Framework*, will help us discover the causal mindset, a five-step framework, allowing us to question causal claims in a structured way and decide whether we need further evidence or not.
- *Chapter 6, Randomized Experiment*, focuses on randomized experiments and why it is often called the gold standard in causal inference.
- *Chapter 7, Quasi-Experimental Methods*, will build on the previous chapter. At the end of *Chapter 6*, we will see that randomized experiments have weaknesses and are often impossible to do. This chapter will present another family of solutions, quasi-experimental methods, methods that can fill the gap when randomized experiments are impossible.

Finally, we will come to the final chapters, *Chapters 8* and *9*, which address how to deal with imperfection.

- *Chapter 8, Correct Answer, Wrong Question: The Importance of Choosing the Right Metrics*, will take an unexpected turn and reveal how perfectly sound causal analysis can be misleading. This relates to a famous issue in statistics: type III errors. Type III errors mean that you correctly answer the wrong questions. We will see the source of these issues and how the choice of measurement can solve the problem, as well as discussing Goodhart's law: *"When a measure becomes a target, it ceases to be a good measure."*
- *Chapter 9, Embracing Uncertainty*, will wrap things up and invite you to take the next steps in this journey to understand causality and make better decisions.

To get the most out of this book

You will practice with scientific articles, newspaper articles, and business examples, applying causal inference to varied topics: pollution and respiratory disease, marketing campaigns, weapons and wars, alcohol and accidents, education and wages, meat production and environmental impact, class size and academic success, and many more. Note that in every chapter, I will include an example about measuring the impact of AI, not only as it is a timely and valuable matter for organizations and society but also because it is what I focus on today. Moreover, at the end of each chapter, we are going to explore three case studies. As we advance through this book, we will be able to apply the tools from each chapter to these concrete examples to illustrate the applicability and facilitate learning by practice, demonstrating how each tool builds upon the previous one.

The three case studies will reflect diverse topics and types of applications:

- Case study 1 is about health and daily life experiences: What is the effect of cold showers on health? This case study will illustrate how we can apply such tools in our everyday discussions, but also to better understand scientific research.
- Case study 2 focuses on marketing and policy-making: What has the effect of the energy sobriety campaign in Switzerland in 2022 been on energy consumption? The second case study will allow us to explore the use of causal inference concepts for marketing and strategic decisions, providing insights that are crucial to business leaders and decision-makers.
- Case study 3 will take us into the world of pollution and health: What is the effect of air pollution on mortality? This third case study relates to the growing concerns about the environmental, social, and health impacts of our economic activities. This study applies causal inference to a pressing global issue, highlighting its importance in designing effective environmental policies.

Let us challenge the status quo

From the personal to the global, from the theoretical to the practical, this book will lay a foundation that will allow you to challenge the status quo, armed with the power of a clear, causal understanding. Whether you are making business decisions, shaping public policy, or simply trying to understand the forces that shape our world, the insights from the following chapters will provide you with a new lens to analyze the world. Prepare for this journey full of ups and downs, where we will emerge with the tools necessary to make informed, confident decisions.

Conventions used

Warnings or important notes appear like this.

Tips and tricks appear like this.

Get in touch

Feedback from our readers is always welcome.

General feedback: If you have questions about any aspect of this book or have any general feedback, please email us at `customercare@packt.com` and mention the book's title in the subject of your message.

Errata: Although we have taken every care to ensure the accuracy of our content, mistakes do happen. If you have found a mistake in this book, we would be grateful if you reported this to us. Please visit `http://www.packt.com/submit-errata`, click **Submit Errata**, and fill in the form.

Piracy: If you come across any illegal copies of our works in any form on the internet, we would be grateful if you would provide us with the location address or website name. Please contact us at `copyright@packt.com` with a link to the material.

If you are interested in becoming an author: If there is a topic that you have expertise in and you are interested in either writing or contributing to a book, please visit `http://authors.packt.com/`.

Free benefits with your book

This book comes with free benefits to support your learning. Activate them now for instant access (see the "*How to Unlock*" section for instructions).

Here's a quick overview of what you can instantly unlock with your purchase:

DRM-Free PDF Version

Download DRM-free PDF and ePub copies of this book.

7-Day Packt Library Access

Get 7-day unlimited access to 8,000+ books and videos. No credit card required.

Available for first-time Packt+ trial users only.

Next-Gen Reader Access

Read this book on Packt Reader with progress sync, dark mode and note-taking.

How to Unlock

Scan the QR code (or go to `packtpub.com/unlock`). Search for this book by name, confirm the edition, and then follow the steps on the page.

UNLOCK NOW

Note: Keep your invoice handy. Purchases made directly from Packt don't require one

Share your thoughts

Once you've read *The Causal Mindset Handbook*, we'd love to hear your thoughts! Scan the QR code below to go straight to the Amazon review page for this book and share your feedback.

`https://packt.link/r/1806117851`

Your review is important to us and the tech community and will help us make sure we're delivering excellent quality content.

Part 1

Why Causality?

In this first part, we will quickly climb the peak of enthusiasm. The first part provides the foundational knowledge, allowing us to understand what causality is, why it matters, and how it differs from prediction, which is what we often do with AI. Along the way, you will learn how to build directed graphs to represent causal relationships, which should be the starting point of any causal model. By the end of this part, you will know how to represent causal pathways graphically, how to choose between causal and predictive inference, how to combine them efficiently to measure the impact of AI models as top tech leaders are doing, and what the emerging field of Causal AI is about.

This part of the book includes the following chapters:

- *Chapter 1, Understanding Causality and Its Importance*
- *Chapter 2, Causation versus Prediction*

1

Understanding Causality and Its Importance

Imagine what would happen if you did not understand causality. You would touch a hot pan, and even after being burned, you might do it again. People would be crossing the road without anticipating that the approaching car might not be able to brake, and that the inevitable collision could cost them their lives. Do you think that this sounds crazy? Well, surprisingly many decisions are actually made without understanding causality, by people who ignore the science of causal inference, leading to costly mistakes. Understanding causality is essential to planning, anticipating, and being forward-looking.

This chapter builds the essential foundation for your causal mindset. We will start by exploring a crucial question: how do we actually know the things we believe to be true? This will lead us to distinguish between weak evidence, simple correlation, and the clues that point toward a true causal link.

The ultimate goal of this chapter is to allow you to understand the big picture and clearly understand what causality is, why it is useful, and how to represent it with a diagram. Mastering these foundational concepts is critical because every other chapter in this book builds upon them. In particular, we will see that drawing a causal graph is usually the first step of causal inference and will be done many times throughout this book.

So, get ready to climb and quickly reach the peak of enthusiasm. In this chapter, we will cover the following topics:

- Introducing causality
- Visualizing cause and effect

Introducing causality

I became obsessed with causal inference because I saw the same pattern across public policy, business, and life: people were making big decisions based on opinions and correlations, and not causal evidence. Take the example of AI. Investments in AI are massive, often driven just by the **fear of missing out** (**FOMO**), and many organizations still can't answer the fundamental question: *Is it actually working?*. Surprisingly, there is a way to reliably answer this type of question that is often ignored. It is called **causal inference**, and this book will arm you with the tools to understand it, better question the world, and sharpen your decision-making by separating strong causal evidence from weaker correlational claims.

Of course, this goes way beyond assessing the impact of AI, which is today one of my main focuses. Most of the questions we ask ourselves every day share a common theme: they concern causal relationships.

What is driving climate change? Will stopping the sale of fuel cars reduce emissions and help prevent further climate change? Does sending developmental aid help poorer countries or not? What have been the effects of COVID lockdowns? What is the impact of AI on our jobs? How does quitting a job impact one's career, financial stability, and happiness? Should I stop eating pizza to lose weight, or will it only make me miserable, being hungry all the time?

This list of varied questions reveals how understanding causality is essential for issues ranging from fighting climate change to pursuing our quest for happiness.

In a discussion between Donald Trump and Elon Musk, Trump said, "*In my last week, we had the best illegal immigration numbers, meaning stopping. It was at the lowest.*" Many people will conclude that the cause was Trump's policies on immigration. However, Trump's presidency ended in January 2021, in the middle of the COVID-19 pandemic and after a year of global lockdown. So, what was the cause of this reduction? Trump's political choices, COVID-19, both, or even something else? Understanding the true cause is central to making better decisions and avoiding misinformation. This book will help you shed light on such questions.

For many years, I have been obsessed with these questions, most importantly, with how we can identify a cause reliably. I remember situations from my childhood where this would haunt me, and later, when the teacher claimed causal links, I would always wonder *How do they know?*. Long story short, they didn't most of the time. It is no surprise that one of my favorite citations as a child was from a French comedian, Michel Colucci (also known as Coluche): "*Just because many of them are wrong, doesn't make them right.*"

One day, at the beginning of my bachelor's studies, I was exposed to this concept of causal inference. Something clicked. I was hooked. I would rush home with excitement to read more books about it.

Since then, I dedicated countless hours, days, nights, and years to study, teach, and work with causal inference. I obtained a Ph.D and used these tools to answer complex causal questions: What is the effect of sending weapons on the probability of conflict in Africa (see Gallea, Q., 2023, *Weapons and war: The effect of arms transfers on internal conflict*)? What is the effect of COVID lockdowns on the spread of the virus (see Bonardi et al., 2023, *Managing Pandemics: How to Contain COVID-19 Through Internal and External Lockdowns and Their Release*)? The answers to those questions have been published in top peer-reviewed scientific journals and discussed with policy-makers.

This experience made me realize how causal inference is necessary to decode the world and make educated decisions, but also how these concepts are widely unknown, ignored, or misunderstood.

Hence, I decided to spread this knowledge, making it accessible without losing the rigor. I have taught statistics and causal inference to about 15,000 people, ranging from undergraduates in economics to experienced medical researchers, and including C-suites, or data scientists at large companies.

But it was not enough. Every day, I am still witnessing abusive causal claims, misinformation in the news, in political debates, even in published scientific papers, and on social media. This is why I wrote this book to hopefully make this knowledge more accessible, disrupt the status quo, and empower people to make better decisions.

Let us pause for a moment and examine some of the simplest beliefs we hold. For instance, how do we know that gravity is real? Most of us would say we learned about it in school, that objects fall when dropped, or that the planets orbit the sun due to a gravitational pull. Yet, for most of us, vaguely remembering a teacher's explanation or reading about gravity in a textbook is still someone else's evidence. Ultimately, to truly convince ourselves, we either must collect data on our own, by running experiments (dropping objects, measuring their acceleration, etc.), or we have to trust a process we believe is reliable (such as scientific research). And this should extend to every belief we hold, from everyday knowledge to the most complex questions about our planet, our bodies, or our society.

Take something as mundane as boiling water for tea. How do we know that putting the lid on causes the water to boil faster? Possibly because we have tried both ways and observed that steam accumulates faster, or our parents told us it was true. But if we never tested it ourselves, we might just be following tradition. We assume it is "obvious," but is it truly? The moment we become aware of how often we accept things without any verification, we realize how much of our "knowledge" relies on secondhand, unverified information.

Consider a less trivial example: how do we know that a certain household cleaner actually kills 99.9% of germs, as its label claims? We trust the brand, the advertisement, or the regulatory body behind the scenes. But did we personally run or witness any experiments with Petri dishes and bacterial cultures?

Most of the time, we are comfortable trusting what we hear from authority figures, friends, or family. Yet, history is filled with examples of "certainties" that later proved incorrect. Medical science once believed bloodletting was beneficial for many health issues, from headaches to fevers. It took systematic observation, physicians who tracked patient outcomes, tests comparing those who bled to those who did not, to reveal that the practice frequently did more harm than good.

Even more humbling, entire civilizations built monumental structures to appease gods they believed caused good harvests. Drought or bad weather would be interpreted as a sign that more sacrifices were needed. But what if the real cause was simply the periodic climate variation tied to ocean currents? For centuries, no one systematically measured weather patterns precisely, so there was no data to correct that belief. In many ways, our ancestors were certain, and that certainty was extremely costly.

Today, we like to believe we are far more advanced than those who practiced bloodletting or prayed to the rain gods. Indeed, we have more data and technology available. However, if we examine our most routine decisions, we find many are still based on little more than untested assumptions or traditions:

- **Diet fads**: Why do so many people cut out entire food groups overnight (e.g., carbs or fats) simply because a friend swears by it? *I know it works*, they say, but have they systematically tested it, or did they observe a coincidental drop in weight while simultaneously starting a new exercise routine?
- **Workplace practices**: How often does a manager implement a "best practice" from a business article without checking whether it has robust, causal evidence behind it? Maybe it worked in one high-tech company for reasons that do not apply to a small marketing agency, or was just cherry-picked.
- **Tech gadgets**: Even in the world of technology, many of us assume certain phone or laptop brands are "the best" or "last longer" based on brand loyalty or hearsay. Have we confirmed with longevity data from thousands of users, or do we rely on a handful of personal anecdotes?

In each case, there is a gap between what we think we know and what we have verified or, at least, what can be verified by a robust process. Filling that gap requires more than an offhand opinion or a single correlation. It calls for systematic inquiry with data and experimentation, the very process at the heart of causal inference.

Why go through this effort? Because when we do, we unearth the true cause, not just a spurious correlation. This process of using the systematic method presented in this book to push beyond guesses is the foundation of causal inference. It is how, at a larger scale, scientists test new drugs, engineers design safer cars, and policymakers (should) evaluate whether a tax reform is beneficial.

One might wonder, *Why bother scrutinizing everything? Isn't life easier if we just rely on what we've heard and trust experts?*. In many cases, we can and do rely on experts. However, even domain experts are not always experts in causality, and mistakes happen. Hence, the cost of blindly trusting any causal claim can be steep:

- **Health**: Misinformation about a medication's efficacy might lead people to waste money, or worse, delay proper treatment
- **Politics**: Believing unverified claims about how a policy will "fix the economy" or "protect jobs" can lead to laws that have exactly the opposite effect of what was promised
- **Personal finance**: Financial "gurus" might promise quick riches through a particular strategy, but it could also lead people to lose their savings
- **Business decision-making**: A company might adopt a flashy new AI-first strategy, based on anecdotal success stories from competitors, only to see its own sales plummet because the market was not ready and the quality was lower, despite the gain in productivity

In short, failing to question causality can leave us vulnerable to harm, be it financial, physical, or social. Sometimes, a misguided belief only wastes a little of our time. Other times, it can spark events with massive personal or societal repercussions.

This doesn't mean we need to run experiments or do a literature review of reliable causal evidence for every trivial decision. Day-to-day life requires shortcuts. And this book fills these gaps. First, it proposes a structured way to do this efficiently when proper experimentation is not possible or needed. Second, for bigger decisions, those that can have significant consequences, we should use causal inference properly if possible, and *Chapters 5* and *6* will show you the methods for these advanced situations.

Developing this mindset does not mean we become jaded cynics who trust nothing and no one. Rather, it means we become active participants in the discovery of knowledge, ready to ask good questions and be curious about how the world works while accepting and evaluating uncertainty.

But first things first, let's define causality. **Causality** refers to the relationship between cause and effect. It's the idea that one event or action can lead to another event or outcome. In other words, causality is concerned with understanding how things happen and why they happen.

Can you explain causal inference to my grandparents?

This is a question I get asked a lot, and I think that it is a convenient way to illustrate how this topic is a fundamental part of our lives, and yet few people formally study it. So, let me share with you how I would have explained causal inference to my grandparents:

"Any decision you make is based on causal assumptions. Look, you just planted your tomatoes in this corner of the garden because it receives the most sun. You did this because you know that tomatoes need a lot of sun to grow and be tasty. So your action is based on a simple causal chain: sun → tasty tomatoes. For tomatoes, it is common knowledge and quite straightforward.

Moreover, as you have been gardening for decades, you also know and probably observed that if you keep your tomatoes in the same spot year after year, they tend to grow smaller. This is due to another causal effect, namely, the fact that tomatoes will drain the nutrients in the ground if you do not alternate with other crops.

So these causal relationships are essential for you, but also for bigger decisions.

You like nature, and your garden, right? Now, think about climate change. How do we know that humans are responsible for climate change? How do we know which policies have a positive or negative effect?

All these questions are complex and require careful causal analysis, or they might backfire. Without using causal inference, our decisions might not be better than rolling dice, reading the horoscope, or looking at the stars.

So now, let me boil water for tea and bring biscuits, because I think that I observed a positive causal impact on your mood when I do this after lunch."

Through everyday examples, from gardening to business decisions, we've now seen why understanding causal relationships is so fundamental. This intuitive grasp is a great start, but to build our causal mindset, we need a more formal and precise definition. In the next section, we will define causality more rigorously and understand the role of data, domain knowledge, and scientific research in the pursuit of causal evidence.

What is causality?

To answer these questions of cause and effect and measure and identify causal effects, we can rely on a scientific field called causal inference.

Causal inference serves two crucial roles. First, it allows us to assess causality from past events. As we are going to see throughout this book, you can use data and precise methods to measure causal effects. Understanding past events helps gain knowledge of causal relationships. However, most of the time, you need this information to make future decisions.

The second role of causal inference is therefore to help you predict what would happen if you made one choice rather than another. You need to have credible evidence about the effect of a drug to recommend it to someone; you need credible estimates of the potential effect of a marketing campaign, strategic decision, or policy to make informed decisions. By understanding the causal relationships from past data, you can forecast the potential outcomes under certain conditions, allowing better planning and strategy.

Causal thinking requires elaborate cognitive capacities. In advanced situations, causal thinking requires complex imagination (see Pearl, J. and MacKenzie, D., (2018), *The Book of Why: The New Science of Cause and Effect,*). Humans are forward-looking, able to imagine the future consequences of their actions. We are constantly going beyond our own experience and observation of association to infer cause and effect. We can imagine a long chain of consequences for complex scenarios going beyond our personal experience:

What if we stop having personal cars? We might have less congestion, more efficient vehicles, and parking will take up less space in cities. We could see more parks and green zones in cities. These vehicles should be equipped for families with baby seats available or for people in wheelchairs, which might lead to some standardization of the vehicles, and so on.

Every decision we make, whether as individuals, organizations, or governments, is based on the assumption that certain actions will lead to certain outcomes. As individuals, we constantly make decisions because we expect some consequences. We may decide to reduce meat consumption or buy an electric vehicle in order to reduce our carbon footprint. We may decide to go to the gym because we ate a pizza for lunch, hoping to keep our weight stable. We may put a lot of effort and many hours into a job, expecting a raise. We may take one more shot of vodka, thinking that we will dance better. Managers make decisions daily, expecting results for their firms. Those decisions could involve stopping home office work, based on the belief that this will raise productivity, or about changing the home page of their website to increase subscriptions. Finally, governments make decisions hoping to shift their country or region in one direction or another.

However, finding causal evidence for those relationships is not trivial, as we are going to see throughout this book. Moreover, failing to properly assess the causality of a relationship might lead to negative consequences. Let me illustrate this again with some examples.

What if you replace your fuel car with a new electric car? Feeling good about your "green" vehicle, you use it more than your previous car and actually stop using public transportation (this is called a rebound effect). What would be the net environmental impact of this change? Given the fact that the source to charge your car is not necessarily renewable, that disposing of batteries remains an environmental problem, and that you substituted some of the trips that you did previously with more environmentally friendly means of transportation (e.g., walking or public transportation), the net effect of this choice on the environment might be negative.

Let me use another hypothetical situation with a firm to illustrate the cost of failing to properly identify the consequences. The marketing team of a new streaming service made a significant change to their app's home page. They decided to display the most-watched movies instead of the newest releases. One month later, the management observed that the subscription rate was falling and decided to fire the whole marketing team. What if the change took place at the same time another streaming service was launched, or during a particularly sunny month when fewer people stay at home to watch movies? Maybe the marketing team did a fantastic job and prevented an even steeper fall in subscriptions. But without a proper assessment of the causality, concluding anything is dangerous and might be counterproductive.

The big picture

Our expectations about the consequences of an action rely on hypothetical causal chains, based on our experience, something we have read or heard. Those pieces of evidence might be weak or strong; it might be one piece of personal anecdotal evidence, an article we have read on social media, or decades of scientific literature. The weight we give to each of those pieces of evidence should be proportional to its reliability. Unfortunately, humans have only bounded rationality, and we are all subject to biases, which eventually makes us rank those elements wrongly.

Hence, whenever possible, we should favor objective, reliable sources to reduce the risk of drawing the wrong conclusion. That being said, as we are going to see in this book, while science is often the best that we have, it is not bulletproof. Applying the knowledge from this book will help to separate weak from strong evidence, whether it comes from scientific research or from any other source. Second, multiplying the sources and trying to find opposite views helps to have a broader perspective. In particular, obtaining information from multiple sources might help to address one of our worst enemies in this context: ourselves and our confirmation bias. Confirmation bias is a cognitive bias implying that we tend to focus on and better remember elements that confirm our prior beliefs.

So, what can we trust if we cannot even trust ourselves? My answer is very simple: question everything, from your own beliefs to scientific research, including what I wrote in this book. As a former scientific researcher, I will never trust a result or a claim just because I saw it published in a top scientific journal such as *Science* or *Nature*. I will have a look at the article and try to understand the conclusion, the model, the data, and the limitations. Scientific research is performed by humans and is thus imperfect by nature. However, there is no doubt that this is the best and most reliable source we have. Compared to what an influencer on social media would share, my trust in a scientific article published in a decent journal is significantly higher. But that being said, I will question it anyway. Virtually every study has limitations and weaknesses, as we are going to see in the following chapters. Hence, I will remain skeptical and recommend doing so, unless there is a wide range of evidence going in the same direction, found in different contexts by varied reliable sources.

Unfortunately, there are two main issues with this approach. First, questioning scientific articles is not feasible for many people. Second, in numerous situations, the answer may not exist in the literature, there may be no consensus, or we simply do not have the opportunity to go into depth. The good news is that this book intends to provide solutions to these problems. As you will see, the concepts shared in this book will allow you to challenge the results and claims of some scientific articles. Note that most of the time, the scientific paper is not the issue; it is the interpretation by individuals who use the research abusively. In addition, one of the main purposes of the book is to teach you the *Causal Mindset Framework*, a five-step method contained in *Chapter 5* that allows you to challenge causal claims and correlations in real time in a structured way.

Now, if you can question scientific research with the concepts from this book, imagine how you will look at a political debate, business decisions or the news. Relying on our bounded rationality and often our ignorance of causality, numerous people will sell us a story based on simple relationships, to shift our opinion, and to gain votes or support. Misinformation is everywhere. Today, it is even more complex to identify alternative facts, as AI allows anyone to generate high-quality texts in seconds for free. People no longer need a certain level of education or expertise to use sophisticated language and write clearly and convincingly. Hence, this book aims to equip you with actionable tools and empower you against this ocean of information and misinformation.

Many people like certainties. My personal interactions with some journalists and politicians revealed this problem. Often, people think that either you know something with certainty, or you are incompetent/clueless. But most of the time, certainties are either lies or heroic claims made by people who also struggle to accept their own knowledge gaps and uncertainties. I am convinced that teaching statistics since elementary school would help people to deal with the unknown. It's fine to say that you are not 100% sure about something. It doesn't make you incompetent. On the contrary, it reveals that you know enough to know what you don't know, and you are humble enough to accept it and say it out loud (do you remember the Dunning-Kruger effect?). Statistics is the art of dealing with uncertainty, an incredibly humbling and useful skill to navigate the world. If you listen to my talks and read my posts and articles or this book, you'll witness that, most of the time, I'll use phrases such as *it seems, there is a high risk, it tends to, from my experience, anecdotal pieces of evidence,* and so on. In short, I would rarely generalize and say that something is 100% in a certain way. It depends. Some people are bothered by this and will look elsewhere, and others might stay, exchange, and work with me, appreciating this honest position. I appreciate this self-selection process and embrace the fact that not everyone will like this.

But of course, luckily, it is possible to gain confidence in some areas. To do so, we must go beyond opinions, gather data, cross-check, apply causal inference, and eventually rely on scientific research. So, let's see now the role of each of those pieces to reliably answer complex questions.

Dumb data versus our limited perception

On the one hand, data is essentially *dumb* and does not have a sense of causality. On the other hand, we have bounded rationality and suffer from numerous biases. The confirmation bias mentioned earlier is just one example. Other biases include attentional bias, which is the fact that we tend to focus on certain elements over others. For example, after you buy a new car, you suddenly start noticing that same model everywhere on the road. This focus might lead us to overestimate the frequency of an event while underestimating others.

Moreover, our own experience is limited, and by basing our knowledge solely on this, we would have a rather narrow view of the world. But our capacity to expand our knowledge by learning about the experiences of others might sometimes backfire. Have you already heard that high school dropouts do better in life (e.g., Bill Gates or Steve Jobs) or that paraplegic people often walk again despite a diagnosis telling them otherwise? This is an example of availability bias, where our perception of reality is distorted because we rely heavily on easily available information. Stories about successful high school dropouts or people walking again after a terrible accident are more frequently relayed, while others will have very little attention.

Another interesting bias I want to add here is apophenia, a psychological bias where we tend to perceive meaningful patterns between unrelated things. If we produce a scatter plot with random dots, as illustrated in *Figure 1.1*, we tend to guess trends, and different people might see different things, as it is rather subjective.

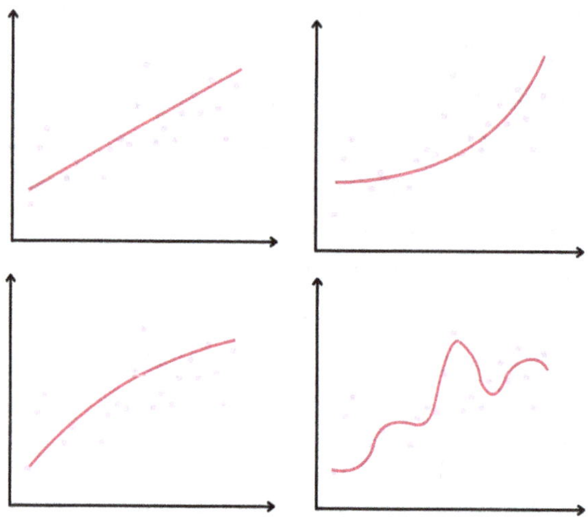

Figure 1.1: Without proper analysis, we tend to imagine a relationship even if it is completely random

Hence, despite choosing between data and our knowledge, we should rely on both jointly, each filling the limitations of the other. Using data allows us to enrich our view and go beyond our limited personal experiences. Furthermore, what we refer to as domain knowledge gives us the ability to critically evaluate and question the relationships observed in the data. Suppose you observe that across countries in Africa, the amount of sun exposure is surprisingly negatively correlated with the number of installed solar photovoltaic panels. Your domain knowledge may help you realize that photovoltaic cells are less efficient when the temperature is too high. Hence, this fact might partly explain the previously described negative correlation. Moreover, income might also play a role, as hotter regions may also have lower average incomes and hence lower resources to invest in the energy transition.

So, we are faced with a dilemma: data is dumb, and our own perception is biased. As we are going to see throughout this book, good causal inference requires both domain and technical knowledge to ensure that we address these weaknesses.

One of the first steps of causal inference is to represent the causal pathway. Formalizing this representation and actually drawing the relationships between the different pieces will prove to be a central piece in our journey. To do so, we are now going to discover directed graphs.

Visualizing cause and effect

We can draw causal relationships with a graph. The arrows within the graph represent the direction of the effect relating two different elements of the graph. In the field of causal inference, those graphs are called directed graphs. The name is self-explanatory. Directed graphs are graphs with arrows representing the direction of the causal relationship. For example, if the rain causes the ground to be wet, you can represent this by *rain → wet ground*.

Advanced technical detail: acyclicity

A popular model to represent causal relationships, called a **Directed Acyclic Graph (DAG)**, is widely used in literature and in the field of causal inference. DAGs have an additional assumption called acyclicity, which prevents cycles or loops. A cycle is a situation where you could go back to a node by following the arrows, hence creating a loop. The assumption of acyclicity forbids this and, in particular, the use of bidirectional arrows. The purpose of this assumption is to simplify the situation. Without cycles, it is easy to identify the before-and-after relationships, allowing us to work conceptually or mathematically with such a model more easily. However, relaxing the assumption of acyclicity allows for the depiction of mutual relationships that are often encountered in real life.

For example, alcohol consumption might cause depression, while people might deal with depression by drinking alcohol. Thus, this relationship can be simply represented by *alcohol consumption ↔ depression*. This mutual relationship could actually be decomposed into a chain of causal effects once we take into account the timing: *alcohol consumption in t-1 → depression in t → alcohol consumption in t+1* (with *t* representing the time period). You can refer to Reisach et al., *The Case for Time in Causal DAGs* for an in-depth discussion. However, in practice, we can often conceptually decompose using time, but the data restricts this capacity. During my Ph.D., I studied the effect of weapons on wars. While this mutual relationship is not instantaneous, I worked with yearly data, and within a year, it was valid to assume that violence increased weapon' imports, which increased violence. For these reasons, in this book, I will not constrain my graphs with acyclicity and will just use directed graphs.

This simple visual tool will allow us to exemplify one of the defining principles in causal inference: why correlation does not imply causation. And indeed, a correlation is an observation that two things move together, not necessarily in a causal way. In this book, I use the terms association and correlation interchangeably to refer to statistical relationships between variables for which causality has not been established. Note that the relationship could be positive, negative, or more complex (non-linear), as represented in the following graph:

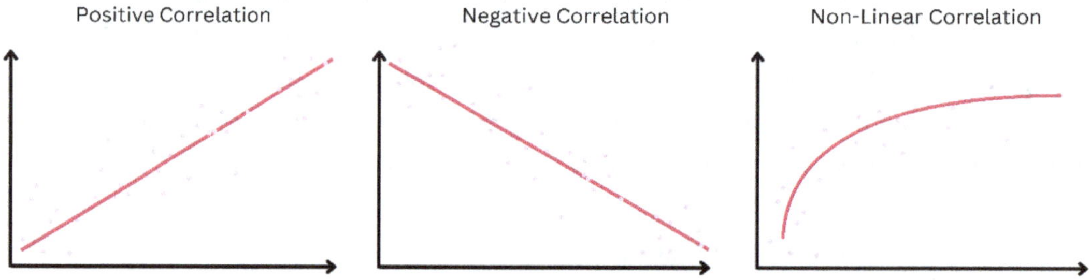

Figure 1.2: Visual representation of graphs with positive, negative, and non-linear correlation

While many people know that correlation does not imply causation, very few understand why. There are two main scenarios. First, as illustrated in *Case 1* of *Figure 1.3*, the positive relationship between shark attacks and ice cream sales is just due to a common cause: the weather. When it is sunny, both take place, but there is no direct causal link between the two. This is what we call a **spurious correlation**. The second scenario is depicted in *Case 2*. There is a direct effect of education on performance on the job, but cognitive capacity affects both. So, in this situation, the positive correlation between education and job performance is confounded with the effect of cognitive capacity.

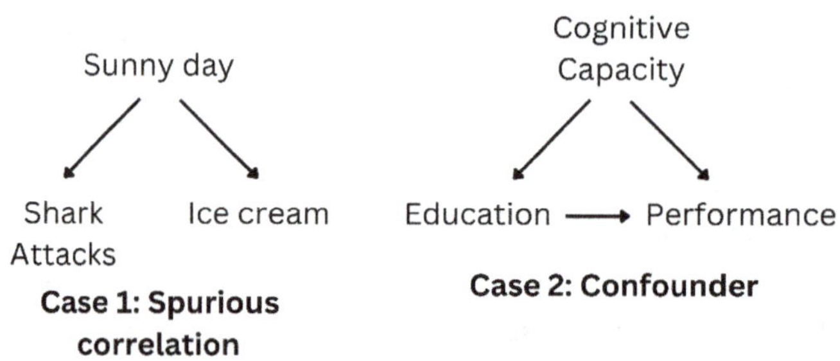

Figure 1.3: Simple examples of spurious correlations and confounders with directed causal graphs

Sometimes, you even find a correlation between different elements just by coincidence. A famous example illustrated on a website dedicated to those curious correlations is the correlation between the number of people who drowned by falling into a pool and the films Nicolas Cage appeared in each year (this example, along with other numerous ones, is available here: `https://www.tylervigen.com/spurious-correlations`).

This difference between correlation and causation is fundamental; it is at the core of a lot of misleading information and numerous costly mistakes. The rest of this book will focus on solving this problem and finding how to separate the two.

It is important to note that causal links can be direct or indirect. Think about the effect of environmental activists blocking roads. They aim to raise environmental awareness, reduce fossil fuel usage, and, ultimately, reduce pollution. However, those effects are indirect. The direct effect of the roadblock is to increase fossil fuel pollution due to the traffic jam created by the roadblock. The two effects are represented graphically in the following figure. Note that this is just a representation, and the causal effect of roadblocks on pollution should be studied thoroughly to prove such effects, and assess the direction (positive/negative) and the magnitude.

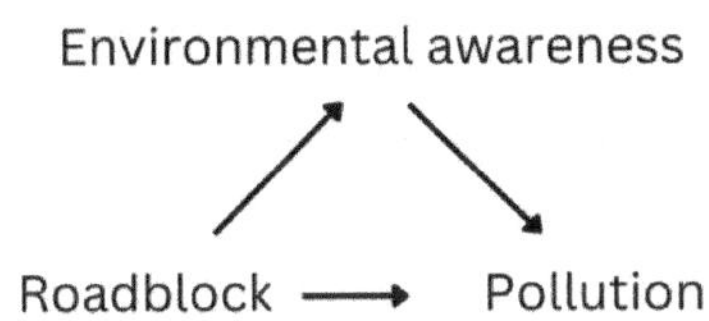

Figure 1.4: Directed graph representing the impact of environmental activists blocking roads on pollution

Technically speaking, an element, a node, connecting the cause and the consequence indirectly, is called a **mediator**. So, in this example, environmental awareness is a mediator. This is the intended pathway, as environmental activists use roadblocks specifically to raise public awareness. A mediator explains how the cause affects the consequence. Hence, it is about identifying the pathway or channel explaining the causal effect.

Most of the time, we are interested in the total effect, which is the sum of the direct and indirect effects. In the example of roadblocks by environmentalists, the hope is that the indirect negative effect on pollution through increased environmental awareness surpasses the direct increase in pollution caused by the traffic jam.

Let me take another example in the business world. Today, AI is everywhere with promises to boost productivity. While it might indeed boost productivity if you use it well, it might also make you lazier and over-reliant on these tools. One study found surprising results when testing the impact of different generative AI tools on HR recruiters. In these randomized experiments, some recruiters had no access to AI, while others received AI assistance with varying qualities. This paper by a Harvard Business School post-doc, Fabrizio Dell'Acqua, titled *Falling Asleep at the Wheel*, revealed that humans with higher quality AI assistance did worse than those with lower quality AI (see Dell'Acqua, *Falling Asleep at the Wheel: Human/AI Collaboration in a Field Experiment on HR*

Recruiters) . The author managed to track the effort of the recruiter to reach their decisions: time spent and the number of clicks (reflecting the number of sections of a candidate's application each recruiter clicked on to uncover). AI might often give good advice, which might positively affect the output quality. However, using AI, and in particular, depending on its quality, also affects the human effort, which subsequently also affects the final output. Following the paper title, these findings suggest that better AI help might negatively affect the joint AI plus human output, as humans might fall asleep at the wheel.

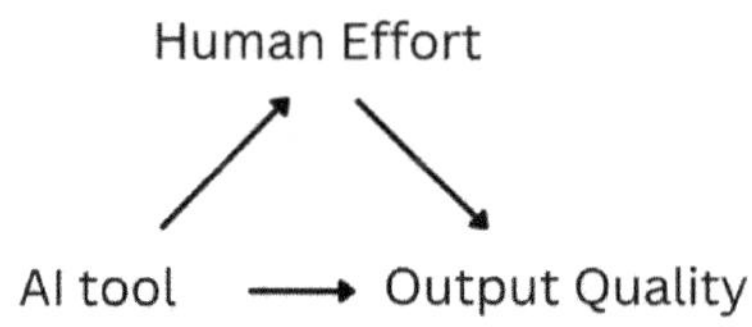

Figure 1.5: This directed graph represents the impact of using AI on output, as studied in the paper Falling Asleep at the Wheel: Human/AI Collaboration in a Field Experiment on HR Recruiters

The distinction between direct and indirect causes raises profound questions about causality: what is the true root cause, and where do we draw the line?. Should we start every causal graph with the Big Bang as the root cause, followed by the evolution of life, leading us to the Industrial Revolution, the use of fossil fuel, and eventually, environmental activists gluing themselves to the road? Theoretically, we could, yes. However, it is obviously counterproductive as one of the aims of such graphs is to visualize causal pathways in a simple, tractable way. Moreover, the idea is to understand the effect of something at one point in time, while some past events have already taken place. Therefore, we aim to limit our graphs to the essential elements.

So, where do we start? We will start with our root cause of interest and ultimate consequence: here, roadblocks by environmental activists and the effect on pollution. We observe the situation at one point in time and want to assess at this moment what the effect of roadblocks is. Hence, we can discard previous causes (e.g., the Industrial Revolution) from the graph, or further consequences that are not of direct interest or of minor importance (e.g., people leaving work earlier, anticipating potential roadblocks) unless these elements are directly affecting the cause and the outcome.

Indeed, it is essential to include the other factors affecting our two elements of interest (root cause and consequence) in the graph. For example, the price of renewable energy arguably affects both pollution and the likelihood of actions by environmental activists. For the sake of the example, let's assume that the cheaper renewable energies are, the more likely environmental activists are to act, as they know that there are accessible alternatives to fossil fuels. We can represent these relationships as follows with a directed graph:

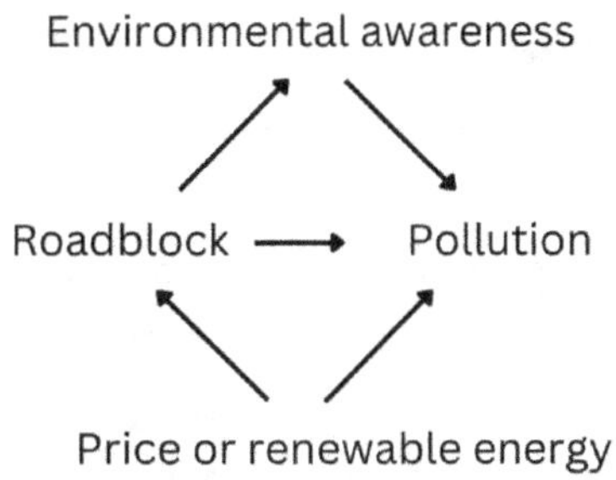

Figure 1.6: This directed graph represents a slightly enriched version of the previous graph

This kind of additional factor affects the relationship between our cause and consequence. An element simultaneously affecting the cause and consequence is called a **confounder**. As we are going to see in the next chapters, taking into account such elements is central for distinguishing a simple correlation from a causal effect. How do we know when we are done with our directed graph? We are never really "done." These graphs will constantly evolve, depending on new evidence and additional knowledge. However, if you follow the rule described previously, you will have your first draft. Despite their imperfections, visualizing our assumptions and understanding of a causal phenomenon on paper is incredibly useful. This is necessary to draw causal conclusions, and hence, not doing it is just a way to hide potential loopholes or weaknesses in your analysis. In addition, it should be clear why I argued earlier that domain knowledge is key. Domain knowledge is what allows you to think about all the potential relevant factors that you have to take into account in order to draw a valid graph. Interestingly, what we leave out of the graph is just as important as what we include. These exclusions also result from deliberate choices and reflections.

This is just a brief introduction to causal graphs. *Chapter 4* will dive into this topic in depth, present other types of relationships (e.g., moderator and collider), and explain how to use these to challenge causal claims. However, we need first a few more pieces presented in the next chapters to fully embrace and benefit from this type of analysis. Once we start making these graphs, we will realize that, arguably, everything is multifactorial. To finish this chapter, let us dive into this concept.

Everything is multifactorial

Everything tends to have multiple causes, and identifying one does not necessarily discard the others. This concept might sound trivial, but it is one of the most profound when we think about causality. Often, people debate the cause of a war or a major event, failing to realize that both sides might be right, as multiple causes can coexist. Recently, the author of a popular book about the effect of mobile phones on depression among young people, Jonathan Haidt, author of *The Anxious Generation*, explained that the decline of mental health starting around 2010 was not due to the 2008 financial crisis but extensive use of mobile phones and social media. Both could be contributing factors, and suggesting one does not invalidate the other. Causal inference is actually key to help identify the relative importance of different causes to shed light on such questions.

When preparing for my TEDx talk, the host asked how I would introduce causal inference to a child. I answered that I would tell them that everything is multifactorial. This simple fact provides perspective and deepens our understanding of both our behavior and the world. Even in a discussion with my young daughter, who was frustrated by a broken pen at school, I found this concept useful. Once, my daughter's teacher told me that she cried and was completely desperate because her friend broke her pen. Is it a tantrum? Should I just say that she should stop crying for something so insignificant? I do not think so. It was a pen she really liked, having just received it for her birthday. Moreover, it was not the first time that her friend had broken her stuff. Finally, it was a day on which her little sister was sick, and nobody had slept well. Understanding that her sadness was potentially an accumulation of causes helps to understand the problem more deeply, better react by providing some perspectives, and potentially fix the issue by reacting adequately.

Now, think about an arguably bigger question: why is the population of South Korea falling? In 2020, the fertility rate in Korea was the lowest in the world, around 0.8. This situation is most likely due to numerous factors, from cultural to economic and social. Identifying and studying these connections is necessary to shape policy and make good decisions.

Beyond these simple examples, it is often insufficient to simply accept that each cause might exist. Indeed, you often want to precisely determine the effect of each cause and what would happen if one did not exist or if its impact were to increase. If you have an ad campaign, you don't want to just say that seasonality, competition behavior, and your ad campaign affect your sales. You want to compute the return on investment of the ad campaign precisely by disentangling the other effects. Acknowledging that many factors are at play is the first step. The tools of causal inference, which we'll explore in this book, help us disentangle these multiple influences, estimate the relative importance of different causes, or isolate the effect of one specific factor even when others are present.

Case studies

We will now explore our case studies in more depth and begin applying the initial concepts to each of these scenarios. Take a piece of paper, or open a document, and try to answer before reading the solution.

Case study 1: Cold showers and the immune system

You or your relatives might have been caught by the cold shower or cold water exposure trend. If you or people you know took cold showers in the past, use your examples to reflect in this part of our case study. If you do not know anybody, just use your imagination. Note that the main rationale for cold exposure is that arguably it might boost the immune system.

Read the following conversion and try to identify where and how some biases might affect the perception:

Mila: Hey, Trent, how's it going?

Trent: Awesome, thanks. Sneeze

Mila: Bless you, caught a cold? It's going around the office with this cold wave.

Trent: Oh, just something tickled my nose. I've been taking cold showers for a month now—never felt more alive. Haven't been sick once, actually.

Mila: Really? Cold showers? How does that work?

Trent: Sneezes again. Just end my showers with a minute of cold water. Scientific research has shown that it boosts your immune system.

Mila: Sure you're not just coming down with something? You seem a bit off.

Trent: No way, I feel great. It's all those cold showers. Plus, tons of people are doing this today, and they say the same. We're like immune to colds now.

Mila: That's interesting. I've read mixed things about that.

In this short example of a daily life coffee-break conversation, several biases play a role. Let me mention two biases that might influence the perceived causal effect of cold exposure on health:

- **Confirmation bias**: Trent might be experiencing confirmation bias, where he attributes his feeling of vitality and lack of illness to cold showers, ignoring other contributing factors such as changes in diet, season, or even placebo effects. This bias might also affect the way he selectively remembers and uses evidence or even refuses to accept facts (e.g., he might have a cold). For example, Trent might ignore other research that might find mixed evidence about the effect of cold exposure on health.
- **Availability bias**: The voices of those who have positive experiences with cold showers are likely louder and more prevalent in discussions, both online and in person. The fact that people who practice and are convinced by the effect are more vocal might amplify the impression of a positive effect.

Finally, I think that it is important again to refer to uncertainty and statistics. Even if cold exposure boosts the immune system, it does not shield you 100% against any disease. So, the fact that the person probably has a cold is not enough to invalidate the argument either.

Now, try to draw your causal graph to study the effect of cold showers on health. For this exercise, we are just exploring potential links. Each link should then be confirmed with a rationale and proof (ideally, scientific causal proof). For the sake of the exercise and parsimony, as this book is not a medical research on cold exposure, I will just outline a potential rationale for each link.

Cold showers might directly affect your health by boosting the immune system. In addition, taking cold showers might affect your perceived health, which then indirectly impacts your health (e.g., placebo effect).

Your health might affect the probability of taking a cold shower. People who are healthier and stronger might be more inclined, on average, to go through the potentially painful process of cold showers. But also, if you are suffering from some health issues, you might be willing to use cold showers to address this problem (e.g., some people aim to use cold showers to reduce chronic pain).

Then, someone who cares about their health might have several healthy behaviors, which increases the probability that they take cold showers and that they are healthy. In addition, there might be some seasonality playing a role as well. When it is winter, and it is cold, people might be less likely to take cold showers, but also tend to get sicker (e.g., common cold or flu).

Lastly, external factors such as 'others' opinions or public knowledge about cold showers might affect your probability of taking a cold shower.

This has been depicted in the following figure:

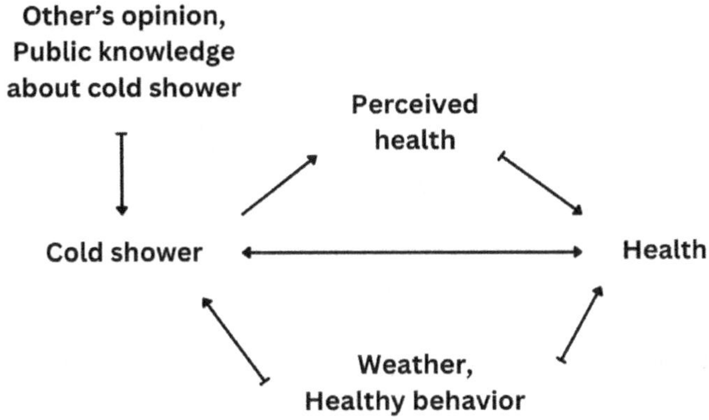

Figure 1.7: Directed graph representing the causal relationship between a cold shower and health

Case study 2: Marketing campaign for energy sobriety

In September 2022, Switzerland implemented an advertising campaign in favor of energy sobriety. This national campaign (composed mostly of ads in journals, street signs, public transportation, TV, or online) presented tips to save energy (e.g., turning down the heat and wearing a sweater, putting the lid on when you boil water, etc.). This campaign was a reaction to the threat of an energy shortage due to the war in Ukraine. It is important to note that it is cold during the winter in Switzerland, and heating systems are a big part of the energy consumption during this period. The campaign started in September.

Now, try to draw your causal graph to study this campaign for energy sobriety on national energy consumption.

The war in Ukraine was the trigger for the energy sobriety campaign. This campaign aimed to reduce energy consumption. The energy sobriety campaign started in the fall when the temperature started going down in Switzerland. The temperature affects energy consumption as colder weather means more heating. Finally, with the fear of energy scarcity in Europe at the time, caused by the war, the price was expected to rise. Higher prices would reduce energy consumption.

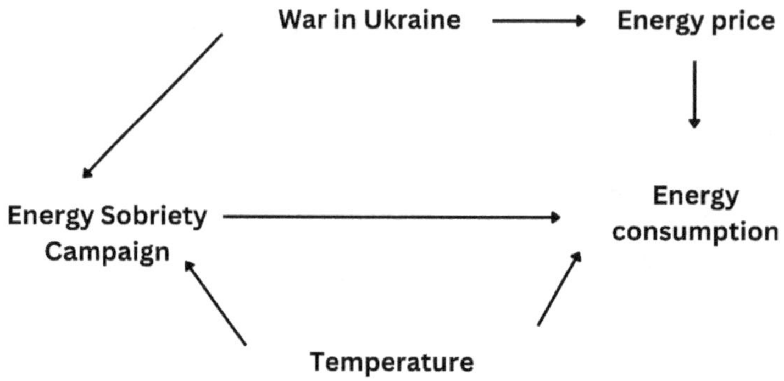

Figure 1.8: Directed graph representing the causal relationship energy sobriety campaign and energy consumption

Case study 3: Air pollution and respiratory disease

Pollution is not only affecting the climate but also directly affecting our health. Recent research estimated that air pollution was responsible for about 8 million deaths per year (See `https://www.bmj.com/content/383/bmj-2023-077784`). As a comparison, deaths from COVID-19 in 2020 were estimated between 1.8 and 3 million depending on the method of computation (estimation from the World Health Organization: see `https://www.who.int/data/stories/the-true-death-toll-of-covid-19-estimating-global-excess-mortality`).

Try to draw the causal graph for this final case study to explore the relationship between air pollution and respiratory diseases.

Air pollution is known to affect health, particularly through respiratory diseases.

Income creates differences with respect to the household location (more or less polluted), healthcare access thanks to financial resources, and education, which might affect health risk awareness and healthy behavior (e.g., probability of smoking).

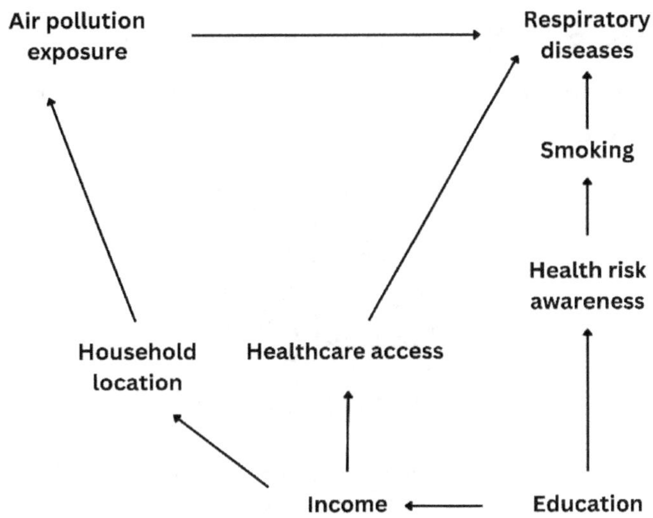

Figure 1.9: Directed graph representing the causal relationship between air pollution and respiratory diseases

Summary

In this chapter, we began our journey into the causal mindset by exploring what causality is and the science behind it: causal inference. You learned that while our world is full of interesting patterns, making sound decisions requires us to dig deeper and ask "why" an outcome occurred. We established that most events are multifactorial, driven by many interconnected causes, and that our own perceptions are often clouded by cognitive biases. To cut through this complexity, you were introduced to your first practical tool: the causal graph, a simple way to visually map out relationships. This foundational knowledge is crucial because it provides the basic language and framework for everything that follows.

Now that we have established what causality is and why it's important, many people might wonder whether we can use AI for this. AI seems to be everywhere nowadays, and many people try to use it to solve any problems. The next chapter will reveal that AI is not the right toolbox to answer causal questions, and that it can even lead to terrible results if you use AI blindly for this, something I witnessed often in my work.

So next, we will explore the critical distinction between prediction and causation, clarifying when to use each approach and why confusing the two is one of the most common and costly mistakes in today's data-driven world.

Get this book's PDF version and more

Scan the QR code (or go to `packtpub.com/unlock`). Search for this book by name, confirm the edition, and then follow the steps on the page.

UNLOCK NOW

Note: Keep your invoice handy. Purchases made directly from Packt don't require an invoice.

2

Causation versus Prediction

> *"If all you have is a hammer, everything looks like a nail" – the law of the instrument, by Abraham Maslow*

This classic saying perfectly illustrates the current landscape of **Artificial Intelligence and Machine Learning (AI/ML)**. Today, ML (a family of models within AI) is in high demand. From research to industry, people tend to focus on those exciting new models to solve every problem.

ML is primarily used for prediction, and it is incredibly good at it. It detects patterns within large and complex datasets. Let's call this type of AI: predictive AI or predictive ML. I will use this terminology to ensure that we are excluding causal AI or causal ML, which I will discuss later in this chapter.

Predictive inference exploits correlations. So, as you now know that *correlation does not imply causation*, you should understand that predictive AI should not be used blindly to measure causal effects. This issue is central to making educated decisions and choosing the right set of tools. I have witnessed even AI experts who didn't understand this difference and used the wrong tools to answer their questions.

This chapter will allow you to understand the differences and why you should not mix predictive and causal inference blindly. Moreover, you will learn how to combine their strength and what the emerging field of causal AI is. By the end of this chapter, you will be able to choose wisely, allowing you to avoid these common mistakes.

In this chapter, we will cover the following topics:

- Why words matter: The thin terminology frontier
- Different roles: Predictive inference (predictive AI) versus causal inference
- Choosing between traditional ML and causal inference
- A powerful collaboration: When causal inference allows us to measure the impact of predictive tools
- Unification: Causal AI, an exciting, emerging new field

Why words matter: The thin terminology frontier

The frontier is often thin. People will make causal claims without realizing it. The terminology and the choice of words are critical in this context. We should clearly refer to non-causal relationships as associations or correlations. When you have a correlation, you can say "When I observe X, I observe Y," for example. This wording does not imply causation. If I say, "When ice cream sales are high, there are more shark attacks," here it is clear, hopefully, that it is a correlation. For causal relationships, you would say: X impacts Y or X increases Y. But we have to be careful. People tend to jump quickly to conclusions. If you say that the crime rate is higher in places with more migrants or foreigners, many people will jump to the conclusion that migrants are the problem. While here, migrants might initially relocate to poorer areas of a country where the crime level is already high. Answering this question causally is complex and requires the use of causal inference. For those interested in a rigorous causal analysis of this topic, a paper published in the high-impact American Economic Review titled *The Violent Legacy of Conflict: Evidence on Asylum Seekers, Crime, and Public Policy in Switzerland* (`https://www.aeaweb.org/articles?id=10.1257/aer.20170263`) offers an excellent example. This journal has particularly high standards when it comes to causal analysis.

This distinction between correlation and causation, or prediction and causation, isn't just a technical detail; it carries massive implications for businesses, governments, and everyone in between. So, let us dive into this subtle but crucial distinction, which has important consequences and is often misunderstood even by experts.

Here is a situation illustrating this issue that I faced recently. Researchers from a leading business school presented to me their research idea. They wanted to pin down which environmental measures and **ESG** (ESG stands for **Environmental, Social, and Governance**, a framework used to evaluate how an organization manages risks and opportunities related to environmental sustainability, social responsibility, and governance practices) metrics are effective in changing behavior, and in particular, reducing the environmental impact of businesses.

I pointed out that their analysis would be correlational. The team quickly said that it was not a problem; we could still use these results. I asked, "How?", "What can we learn from this?". The team was basically stuck because the only useful conclusion would imply causal interpretation. Even if they used the right terminology, people would still probably misinterpret and misuse the results. For example, if they claim that when top executives' bonuses are conditional on the CO2 emissions of the company, the environmental impact is lower, even if they say that it is just a correlation, a vast majority of the audience would read this as a causal link. But this isn't a causal link, or at least, with this study design we can only capture a correlation, not prove causation. It could be that companies that are able to abate emissions set these metrics because they know that they can do it. To measure this causally, we would need to apply one of the solutions presented in *Chapter 6* or *Chapter 7*.

Now, let us see an example of the misinterpretation of prediction versus causation in the news.

Here is another example highlighting that, even when people use the right terminology (e.g., correlation or association), they tend to misinterpret the results. CNN wrote an article titled *6 exercises that address an often overlooked key to longevity*. The article reports a pretty strong correlation observed between grip strength and longevity. Grip strength is indeed a good predictor of heart health. Yes, it might be a good predictor. When I say predictor, think about my point on predictive AI or predictive inference: it relies on correlations. While the article mentions associations and correlations, it does not prevent the author from showing exercises on how to increase your grip strength and use a title implying that it will affect longevity (causal assumption). You can say "correlation/association" all you want in an article, but if you suggest changing the grip strength, it reveals that you didn't understand that it is just a correlation. It is most likely that the main driver of this association is a common cause for both longevity and grip strength, related to physical activity. This is what we call a confounder and what we will study in depth in *Chapter 4*.

Note also that the CNN article refers to a scientific article. But again, if you look at the abstract of the article, even if you are not a researcher but just with the knowledge acquired so far, you would understand that it is correlational due to the terminology used: *"There is strong evidence for an association between lower handgrip strength with higher all-cause, cancer, and cardiovascular mortality risk"* (López-Bueno et al., *Thresholds of Handgrip Strength for All-Cause, Cancer, and Cardiovascular Mortality*, `https://pubmed.ncbi.nlm.nih.gov/36332759/`).

This is an incredibly useful tool to prevent abusive claims based on scientific research. Check whether the research reports an association or correlation, and then interpret adequately. So, next time you see an article or someone referring to research, try to have a quick look at the underlying paper to verify the claim. While imperfect, this will allow you to detect some misleading conclusions in many situations, as was highlighted in this example.

Different roles: Predictive inference (predictive AI) versus causal inference

Predictive AI is all about predicting the value of something by observing the value of other things. This is what we can call predictive inference. Imagine looking outside and seeing people wearing shorts and flip-flops, you'd reasonably predict it's warm. That's predictive inference in action: you observe something (people's clothing choices) and infer the outcome (temperature).

Predictive AI is used widely in varied fields:

- **Spam filtering**: Scanning incoming emails to determine whether they're legitimate or junk
- **Medical diagnostics**: Analyzing MRI images to detect tumors
- **Fraud detection**: Identifying unusual banking activities that hint at fraud
- **Sales Forecasting**: Projecting future sales based on varied economic indicators, and market trends
- **Climate risk**: Observing varied climate and weather indicators to predict the risk of forest fires

One particularly popular type of predictive AI model is recommendation algorithms. The goal is to recommend what a customer might like based on their behavior and the behavior of others. This type of predictive algorithm is at the heart of many big tech companies today. Large streaming companies aim to predict the next TV show you might like; social media platforms in general try to predict the content that will retain you on the platform; top online retailers try to predict what type of clothes you are likely to buy; and so on. All these models rely on predictive AI. The company doesn't need to understand why you might like a pair of jeans or a TV show; they just want to show you the right content.

These applications highlight how powerful predictive models can be at spotting patterns. However, businesses rarely stop at "What's likely to happen?". They want to know "How do we make it happen (or prevent it)?". As Matheus Facure, author of *Causal Inference in Python*, puts it: *"In most situations, businesses don't care purely about prediction. They care about making decisions that bring more customers, increase conversion, decrease churn, increase profitability, and cut costs."*.

That's where causal analysis steps in, because simply predicting an event doesn't explain why it's happening or how to change it. As you now know, causal inference is the scientific method for measuring cause and effect. It's about understanding the impact of something on an outcome. Causal inference aims to understand what would happen to the outcome if we change something, not just if we observe something as it is. Would sales go up if we increased ad spend? Would employee engagement improve if we introduced flexible working hours?

Typical examples where causal inference is essential include the following:

- **Return on Investment (ROI)**: Assessing how much revenue really stems from a specific project or campaign
- **Return on Ad Spend (ROAS)**: Measuring how advertising contributes to sales, factoring in seasonal trends or competitor moves so you don't overstate your campaign's success
- **Pricing**: Estimating the impact of the price on sales (price elasticity) to optimize the strategy
- **Policy evaluation**: Evaluating the effect of new policies, such as remote work rules or executive pay structures, on team performance or employee satisfaction.
- **Identifying growth drivers**: Sorting out whether a boost in sales is triggered by your marketing campaigns or by external factors such as the weather or economic conditions

While predictive and causal inference are both useful, it is important to understand that their goals are different. In predictive inference, the goal is to accurately predict an outcome. Therefore, you devote your effort to ensuring your model can predict the value or category of an outcome on new, unseen data. For example, you want to predict whether a new email should be classified as spam. In the field of predictive inference, there are numerous metrics to judge the quality of a prediction, and their importance varies depending on the context. For instance, some situations demand minimizing false negatives, even if it comes at the cost of more false positives. This could be the case for tumor detection on MRI scans or for fraud detection.

You would rather have a false positive (flagging something harmless) than a false negative (missing something dangerous). An expert can then review the potential positives and filter out the false ones. Conversely, with a spam filter, you could prioritize avoiding false positives. You would rather have some spam slip through (false negatives) than have a critical email incorrectly sent to the spam folder.

Predictive inference is valuable in the same way that correlations are informative. The only issue is if you interpret the results as causal. Let me illustrate this with a real-world example. ML is used in the medical sector to predict the risk of sepsis (Islam et al., *Machine Learning-Based Early Prediction of Sepsis Using Electronic Health Records*, `https://pmc.ncbi.nlm.nih.gov/articles/PMC10488449/`). Sepsis is defined by the WHO as a life-threatening condition that happens when the body's immune system has an extreme response to an infection, causing organ dysfunction. By predicting this risk quickly, antibiotics can be administered in time, or other measures taken, and lives can be saved.

Temperature is usually a key predictor of the risk of sepsis. If you interpret this as a cause, you might be tempted to act on this directly by reducing the temperature of the patient (e.g., external cooling). If anything, this treatment might do more harm than good and will clearly not save the life of the person. The temperature rise is a mere consequence of the sepsis and not a cause.

Here is another real example. Consider a global retailer striving to optimize its email campaigns to boost sales. A typical ML model can predict who's most likely to open an email or make a purchase. However, the real question, the one that drives revenue, is whether sending the email caused the purchase, or the customer would have bought something anyway. To solve this issue, M&S used causal inference. A talk by two employees from a global retailer discussing the use of causal inference is available on YouTube: *Causal Inference in Retail* (`https://www.youtube.com/watch?v=vGdp8x8li1o`).

When you rely solely on predictive models (traditional ML) for your decisions, you often end up in a reactive mode. Why? Because these models "see" the current state of the world, bank account balances, purchase histories, and clickstream data, and predict what's likely to happen next if nothing changes. For example, in banking, a model might spot a pattern suggesting a specific customer is at risk of default. That's useful information, but it doesn't tell you why that customer is vulnerable.

This is exactly why predictive AI leads often to reactive strategies. You wait and observe the signals in the data, just like noticing that someone's temperature is rising, and then you respond to those signals. That can work well for many tasks: flagging fraudulent transactions, triaging medical emergencies, or forecasting sales.

But if your goal is to understand the root of an issue and act on it, you need more. You need causal inference as it helps you move beyond correlation. Instead of only spotting who's likely to default, you start asking: "What are the underlying drivers of this default risk? What if we offered financial coaching or changed the repayment schedule? Would we see fewer defaults?" That's the power of causal analysis: You're not just observing the world; you're figuring out what happens if you act on it. This unlocks proactive strategies, where you can intervene early and focus your actions on the factors that genuinely drive the outcomes you care about.

As Judea Pearl insightfully put it, the difference between prediction and causation is similar to the difference between "seeing" and "doing." Traditional AI helps you see patterns in the data and react accordingly. Causal inference guides you to act on the underlying drivers to eventually affect the outcome.

Don't get me wrong, prediction is still incredibly useful. Let me point to a completely different situation to be sure that this is clear: armed conflict. Pr. Hannes Mueller is a leading figure on this question. He co-created a free online platform, *conflictforecast.org*, which makes public the results of its ML algorithm aimed at predicting conflict risk at the country level. The goal is to help with prevention and decision-making. In a beautifully written piece, *"The Case for Armed Conflict Prediction"*, Hannes wrote: *"Machine learning improves the prediction of events compared to human experts alone (Benjamin et al. 2023), but the question remains whether it is worth the effort if predictions cannot explain what happens. To answer this, it helps to clarify what prediction actually is. Prediction models aim to forecast what will happen next based on patterns in the data. Causal models, by contrast, try to identify the effect of specific variables or interventions on outcomes. Prediction and causal inference models are both just ways to summarize data with the help of statistical methods. But identifying causal parameters is often the harder task because credible causal inference requires strong assumptions and careful identification strategies (Pearl (2009) and Angrist and Pischke (2009))."*

In decision making, the capacity to explain or interpret a decision is often central in order to justify and judge its quality. When it comes to AI, I am sure that you have often heard that these models are black boxes impossible to interpret. This is indeed a limitation and sometimes an issue with complex AI models.

If you take a popular chatbot today, such as ChatGPT, it has apparently more than a trillion parameters. We can easily imagine how this complicates the ability to interpret a result. This comment is also valid for complex predictive AI models.

However, some predictive AI models are called "interpretable." These would be typically simpler models, such as decision trees or linear regression models. In these cases, a human can relatively easily look at the model and understand its parameters and what drives the prediction.

Another term that is used commonly in the world of AI is explainability. A model is explainable if the results can be understood by a human with additional tools or techniques. This is typically useful if you work with complex models such as artificial neural networks.

On the other hand, causal models are in the vast majority of situations interpretable, which is a central and useful feature of these models. It is often fairly easy to understand the model and explain the result.

Choosing between traditional ML and causal inference

Traditional ML and causal inference are both extremely powerful, but they tackle very different problems, as we have seen. The key is to start with this question: "What do I actually want to know or achieve?"

If your main goal is to predict or classify an outcome, such as spotting which customers might default on a loan or deciding whether an email is spam, traditional ML is a perfect fit. It excels at pattern recognition and uses existing data to predict outcomes based on what you observe. You're essentially asking the following:

- Which category does this data point belong to (e.g., spam versus not spam, fraud versus legitimate)?
- What level of outcome should I expect for this data point (e.g., how many units might I sell tomorrow, given the weather or economic conditions?)?

On the other hand, if you want to move beyond "What is happening?" and dive into why it happens, and more importantly, how to influence it, you need causal inference. Instead of stopping at "This customer is likely to default," you investigate "Why is this customer at risk? Which factors (interest rates, financial education, repayment terms) genuinely drive that risk? What if we changed those factors? Would fewer people default?" Causal inference becomes crucial when you want to evaluate the impact of a decision or policy – think ROAS, marketing campaigns, policy changes, or product launches.

It answers questions such as the following:

- Does offering a discount genuinely cause higher sales?
- Would a loyalty program reduce customer churn?
- What's the effect of flexible working hours on employee productivity?

In these scenarios, you're not just predicting an outcome; you're aiming to change it by identifying and acting on the underlying causes.

By clearly distinguishing whether you're aiming to predict or to change, you ensure that you're applying the right tools and setting your organization up for decisions that don't just anticipate the future but also help you shape it. But what if we want to measure the impact, a causal question, of a predictive AI model?

A powerful collaboration: When causal inference allows you to measure the impact of predictive tools

While predictive inference and causal inference serve different purposes, new possibilities emerge when used together. By combining their strengths, we can create systems that not only anticipate outcomes but also guide decisions by evaluating the impact of actions. This integration is especially critical when deploying AI or GenAI models in production, where understanding the real-world effects of these models is at least as important as their predictive accuracy.

At the heart of many AI-driven systems is predictive inference, as you have understood. For example, one of the largest digital travel companies allowing users to book flights, stays, and more, uses predictive inference to power its recommendation system, predicting which hotel or activity a user might prefer based on their browsing history, preferences, and contextual data. This is a classic ML task: building a model that provides the most relevant recommendations to each user.

However, prediction alone is insufficient for making informed business decisions. Once a model is in production, the critical question becomes "What is the effect of this system on key business metrics such as sales, customer satisfaction, or retention?" This is where causal inference takes over.

A senior ML manager in one of the largest digital companies explained in a podcast with Aleksander Molak that predictive models help identify what a user might like, but only causal inference can answer whether implementing a new recommendation system drives meaningful changes in sales or customer conversions (link to the podcast: `https://www.youtube.com/watch?v=xkx1tXLAP-o`). For this, these types of companies rely on randomized experiments (see *Chapter 5*, in the *The causal mindset framework* section) to measure the causal impact of the recommendation system on user behavior.

This interplay between prediction and causation creates a powerful workflow:

1. **Prediction**: Use machine learning to build a model that predicts user preferences or behaviors (e.g., recommending hotels or generating product suggestions).
2. **Deployment**: Implement the model in a real-world environment, where it interacts with users and starts influencing decisions.
3. **Impact evaluation**: Use causal inference methods to measure the model's true impact on **key performance indicators (KPIs)**.

The rise of **Generative AI (GenAI)** models has further underscored the need to combine predictive and causal inference. Many companies invest massively in GenAI and then fail to measure the business impact. Now, this might sound surprising to you, as we know that there is an entire field allowing us to measure this type of effect: causal inference. Indeed, causal inference is useful to measure the impact of the first deployment compared to the situation without AI. In the second phase, it can be used for optimization. Do you have a new version of your chatbot or recommendation algorithm? You can use causal inference to confirm that this new model has led to better business outcomes, and also whether the cost of deploying it is justified based on the benefit estimated. Saying that it works is clearly not enough. When it comes to business decisions, you need to know precisely the impact, and in particular the financial return, to be sure that it is worth the investment. This analysis can go way beyond financial returns. A GenAI chatbot might answer user queries with remarkable accuracy, but how does its deployment impact overall customer satisfaction, resolution times, or trust from the customers? Causal inference methods are necessary to answer these questions.

The distinction and integration between predictive and causal inference are crucial for several reasons:

- **Maximizing ROI**: Businesses need to ensure that the models they deploy create measurable value. By evaluating the causal effects of ML systems, organizations can allocate resources to models and features that drive meaningful improvements.
- **Understanding systemic effects**: Causal inference allows us to optimize beyond short-term outcomes, such as click-through rates, and reveal long-term impacts, such as customer churn caused by over-targeting or clickbait strategies (see *Chapter 8*, in the *Selecting the right metrics to measure what matters about choosing section*)
- **Building trust**: Causal evaluation provides transparency about the effects of AI/ML systems, helping businesses gain trust from stakeholders, regulators, and users.

We have seen that both types of inference serve different purposes and should be used together. So, what is causal AI?

Unification: Causal AI, an exciting, emerging new field

Over the past few years, a new field called causal AI has been growing and attracting more and more attention. It's an elegant blend of traditional causal inference and the more flexible modeling capabilities of ML. But in my view, this new name has also caused some confusion. People see "causal AI" and assume they can just apply ML algorithms to causal problems without any extra thought, and that can lead to trouble, as we have seen, and also as I have witnessed in my advisory work.

The essence of causal AI isn't about blindly mixing ML with causal inference or using the former to measure causes. It's more like applying causal inference principles first, then sprinkling a bit of ML on top to handle certain tasks, such as dealing with high-dimensional data or identifying subgroups that respond differently to a treatment.

There are three main applications of causal AI these days. Let me walk you through these scenarios where causal AI provides a powerful boost over traditional methods.

First, traditional causal inference typically focuses on the **Average Treatment Effect (ATE)**, the average impact of an intervention across the sample. While that's a good starting point, the average can be misleading. Sometimes, a marketing campaign, a policy, or a drug might show strong overall benefits, but if you zoom in, you could find that the effect is driven primarily by one subgroup. The effect could be strong and positive for one group and destructive for another subgroup. This heterogeneity would be hidden if you just look at the average. So, bottom line, you would be making a decision that has a negative effect on a subgroup (e.g., men versus women, certain age ranges, or particular regions). Causal AI helps uncover these differences by estimating the **Heterogeneous Treatment Effect (HTE)** efficiently. In other words, it allows us to reveal whether the same treatment has varying impacts across different subpopulations and to automatically identify these different subgroups in a data-driven way. One popular approach for doing this is the causal forest method, which was developed, among others, by Susan Athey from Stanford University (see her free online course *Machine Learning & Causal Inference: A Short Course* (`https://www.gsb.stanford.edu/faculty-research/labs-initiatives/sil/research/methods/ai-machine-learning/short-course`). This strategy is particularly useful for optimizing and targeting better interventions. Instead of aiming for the average consumer, you can personalize for subgroups, leading to a more efficient strategy.

Let me take an example inspired by a concrete business example that I can't share publicly. The core of this business is an app. The team constantly looks to improve the app and release new features. If you test these new features on your user base, you might see a positive effect. The larger part of the user base might be experienced users, knowing your app well, and they can easily deal with one more new functionality. But at some point, a smaller but very important share of the users is negatively affected by this strategy: new users. Suddenly, you come to the app and it is incredibly complex, so you leave. This is a huge risk and might profoundly question the long-term growth of the app. If you focus just on the average effect, as the experienced users might represent a larger share than your new users, you might not capture this impact easily. Hence, studying the heterogeneity of the effect for subgroups with causal AI would be way more than just a technical fancy tool, but would significantly impact the growth of your business.

The second main application of causal AI I see in my practice is dealing with high-dimensional data, where there are many potential confounders and complex relationships. This is a bit technical to be honest, but let me do my best to give you the idea in an accessible way. When you build a causal inference model, you must specify the relationship between the variables – for example, between the confounders and the outcome. More specifically, you must make some assumptions about what we call the functional form: whether it is linear, quadratic, and so on. This is painful to do and can create some issues. For example, data scientists or researchers could try several models and select the one giving their favorite results, even if there is no objective argumentation to select that one. To avoid this, we can let ML figure out the best functional form in a data-driven way, as ML is very efficient at dealing with high-dimensional data and uncovering complex function forms. One tool to do this is called **double machine learning**.

The third and final main topic that I would put in causal AI is how we can use GenAI to help us with causal questions. GenAI, and in particular, **Large Language Models (LLMs)** such as ChatGPT are surprisingly good at suggesting potential causal pathways and challenging causal claims. There is a growing body of research in this direction. Very early, just when ChatGPT 4 was released in March 2023, I created a free app using this strength. It is the companion app of this book – the Causal Mindset app – and it is freely available at `https://www.thecausalmindset.com/app`. You can share any causal claims, and it will help you challenge these claims. See the following box for an everyday life example.

The Causal Mindset app example

One of the first tests I did with my app was related to my fitness program. I was preparing for a fitness competition, and I wanted to improve on the rower. I asked my coach for advice. He gave me a solid program and asked me to test my full effort for 1 km that day and again in a month to compare my progress. Of course, as I constantly think about causality, my first reaction to him was: "How do I know that the difference would be caused by the program?". I could come up with multiple other factors affecting my performance at the same time. While this is automatic for me, it might not be for many people. So, I created the Causal Mindset app. The app relies on ChatGPT, but I basically gave it some instructions on how to question causality, which is essentially the content of this book.

Here is the result for this example:

My prompt: I followed a fitness program for rowing for a month. I tested today and went faster. Is it thanks to the program?

The Causal Mindset app answer:

Quick feedback – major flaw

The core issue is you've only tested yourself before and after. That's like trying a new rowing technique for a week, seeing your split time improve, and assuming the technique was the reason—without asking whether you'd have improved anyway.

This is a classic "before-after" trap. Without a comparison group (e.g., people like you who didn't follow the program), we don't know if your gain is from:

The program

OR

- Natural improvement (because your body adapts early when doing any consistent training)
- Increased familiarity with the testing process (you might row faster because you're more comfortable with the ergometer test itself)
- Lifestyle changes unrelated to the program (better sleep, diet, motivation spike)

Is there something else (potential confounders)?

Here's what could explain your improvement apart from the program itself:

- **Practice effect:** Repeated testing often leads to better scores. Your brain and body get used to the test format.
- **General physical activity:** Did you increase activity levels outside rowing (e.g., cycling, walking more)?
- **Nutrition/sleep:** Did you start eating healthier or sleeping more during the program?
- **Motivation and effort:** Were you more motivated to perform well because you were expecting to see gains?
- **Placebo effect:** Believing the program would help can actually make you push harder.

Analysis: How serious is this?

Main issue: no control group.

This is huge. Early training improvements often happen because of novice gains. Your body adapts quickly at the start of any program, whether the program itself is "optimal" or not.

This bias amplifies the perceived effect. You might believe 100% of the improvement is from the program when, realistically, only part of it is.

For example:

If you shaved 10 seconds off your 1 km, maybe 4–6 seconds came from actual fitness gains due to the program, and 4–6 seconds from test familiarity, motivation, and lifestyle factors.

But without comparison, this is speculative.

Conclusion

Your improvement is real, but you can't confidently attribute all of it to the program. Most likely, the program contributed, but so did practice effects, motivation, and natural adaptation. To go further:

- Track more data over time (heart rate, RPE, split times)
- Repeat testing after taking a break from structured rowing to see how much fitness you retain
- Compare your progress to others with similar baseline fitness

There is now even the possibility to call this causal knowledge without Python code with a new library called `pywhyllm`. However, note that in any case, with my app or a package such as `pywhyllm`, it is central to always question the output. Generative AI is interesting to brainstorm and provide ideas in this context, but it doesn't know the truth, and it might provide inaccurate suggestions. Hence, as with anything with GenAI, take the time to evaluate the quality of the answer.

As I am writing this book, I can really imagine that the field of causal AI will grow, and more and more of the traditional techniques in causal inference are now augmented with the help of ML. However, as I explained before, causal AI always starts with a traditional causal inference approach and concepts. Hence, the concept of this book won't become obsolete. On the contrary, interest in this field is growing, and the existence of causal AI makes it even more appealing.

Case studies

Let's apply what we've learned by exploring potential applications of predictive and causal inference within our three running case studies.

Case study 1: Cold showers and health

How could predictive and causal inference be used to investigate the claim that cold showers are beneficial for health? What would each approach tell us?

Potential answer

Predictive models could identify whether people who take cold showers tend to have better health. This correlation might be useful for an insurance company trying to predict a person's health level. However, it does not prove causation, as it could simply be that more health-conscious individuals are drawn to taking cold showers.

Causal inference would be required to answer the critical question: "Does taking cold showers actually cause an improvement in health outcomes?"

Case study 2: Energy sobriety campaign in Switzerland

Reflect on potential use cases of predictive and causal inference related to energy consumption in Switzerland and suggest applications.

Potential answer

Predictive inference could be used to detect poorly insulated houses by analyzing thermal images and spotting where heat is escaping. This is a powerful way to flag potential issues for human evaluation, which can then identify homes needing renovation.

Causal inference would answer the subsequent, critical business questions: Does using this predictive model actually reduce energy consumption or lower detection costs? Furthermore, causal inference is essential or evaluating the impact of the separate energy sobriety marketing campaign.

Case study 3: Air pollution and respiratory disease

Suggest applications for both predictive and causal inference in the context of air pollution and respiratory disease.

Potential answer

Predictive modeling can help anticipate pollution increases. For example, an AI-driven system could forecast high-pollution days, automatically triggering alerts and recommending preventive measures such as wearing a mask or avoiding running outside.

Causal inference would then be used to measure the impact of these recommendations on health, mortality, and the prevalence of respiratory diseases. In addition, causal AI could help uncover the heterogeneity of these effects, revealing which policies work best for specific populations or regions. Finally, it would also be essential for understanding the causes of the pollution itself. By identifying the major contributors, policymakers can design interventions to reduce pollution at its source.

As this illustrates, predictive AI enables a reactive strategy (responding to a forecasted risk), while causal inference empowers a proactive strategy (addressing the underlying causes).

Summary

In this chapter, we tackled the critical distinction between prediction and causation, a fundamental concept for making reliable data-driven decisions. You learned that predictive AI is incredibly powerful to predict but relies on correlations and pattern detection, preventing it from understanding the "Why?". Causal inference, on the other hand, is required to answer the more strategic question: "Why is it happening, and what can we *do* about it? What is the impact of this decision or action? Or, what is the cause of this problem?". We explored the important role of terminology and how misinterpreting a prediction as a cause can lead to flawed strategies, using real-world examples from medicine, media, and business. A particularly important concept was that we should combine the two. In particular, if you want to know the business impact of AI tools, you need to use causal inference. Finally, we peeked into the exciting frontier of causal AI, seeing how it combines the strengths of both fields to boost some technical aspects of causal inference.

Now that we understand fully the necessary role of causal inference, we will discover the other, darker side of the coin: proving causality is a difficult task. In the next section of the book, we are going to uncover these challenges before exploring the solutions in the third section.

Join our community on Discord

Join our community's Discord space for discussions with the authors and other readers: `https://discord.gg/G4YmB6DQvf`

Part 2

What Are the Main Challenges?

In this second part, we'll dive into the valley of despair as we discover why causality is actually difficult to prove. Once we understand the main challenge, known as the fundamental problem of causal inference, we will unpack it and clearly identify the main culprits: confounders, reverse causation, and measurement errors. By the end of this part, you will know precisely why correlation does not imply causation and will be able to challenge causal claims by identifying the main barriers to measuring causality.

This part of the book includes the following chapters:

- *Chapter 3, Why Is It So Hard To Prove Causality?*
- *Chapter 4, Beyond Correlation: The Main Culprits*

3

Why Is It So Hard To Prove Causality?

We have now understood, from *Chapter 1, and Chapter 2* why causality is essential for business decisions, everyday life, policymaking, and many other situations. This is fantastic news: we just have to measure things causally to make better decisions!

This is where things get complicated. Actually, as we will study in this chapter, the only way to perfectly measure a causal effect is something impossible. So, buckle up – we are going to dive into the valley of despair. But don't worry, this doesn't mean that we can't provide causal evidence to answer important questions. We will explore what we can actually do at the end of this chapter and, of course, in the next section focusing on the solutions. Long story short, we are going to put ourselves in the shoes of a police detective because that is exactly what it is about: finding the culprit – the true cause. Who doesn't like playing the detective in Cluedo, or following a CSI TV show episode? This is one of the numerous reasons why causal inference is fascinating.

In this chapter, we are going to study the following topics:

- Understanding the fundamental problem of causal inference
- Exploring counterfactuals
- Proving causality

Understanding the fundamental problem of causal inference

When we try to measure causal effects, we face an issue – an issue so central that it is called the fundamental problem of causal inference. To illustrate this concept, let me share with you "the key" – a key you will always be able to use as a reference point. This key will help you unlock the box full of mysteries of causal inference.

Here is the key. It starts with a short story. Imagine that you have a headache. It started slowly, but it is quickly getting more and more unbearable. You massage your temples, look away from the computer, or stop whatever you are doing. You are not sure whether you want to take a pill or not. This is really bothering you because you just bought a book on causal inference, and you really want to keep reading and learning about this fascinating world. So, you wait a bit, drink some tea, and it keeps getting worse. Finally, you cannot take it anymore, so you decide to take a pill.

After a while, you are relieved – the headache is gone.

However, if you are like me, you frown, you squint your eyes, and you start wondering: "Did I get better thanks to the pill? Was it because I drank tea or plenty of fluids? Was it just because time went by?"

In this situation, it is actually impossible to know what the causes were and whether the pill played a role or not. All those potential causes are mixed together. They are confounded. This is true for most situations. The world is complex, and as we saw in the first chapter, virtually anything is multifactorial (has multiple causes).

Hence, ideally, what we want is to isolate the cause and its treatment effect. The only way to know whether the pill was the cause with certainty would require two parallel worlds: exactly identical worlds from the Big Bang to your headache. In one of those worlds, you take the pill, and in the other, you do not. If, and only if, you are doing better in the world where you took the pill, you could claim that the pill was the reason. In this ideal setup, the only difference is with respect to the pill, and hence any difference in the results can be attributed to the difference in treatment. This is depicted in the following figure:

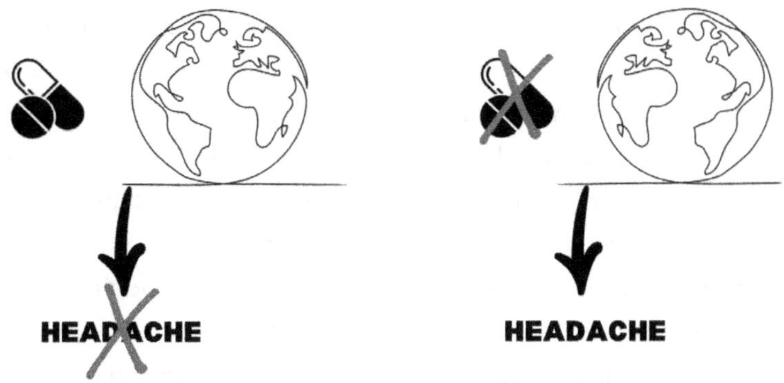

Figure 3.1: The fundamental problem of causal inference illustrated

Unfortunately, we do not have access to parallel worlds to do this. Hence, many things happen at the same time and are confounded (e.g., drinking tea, taking a pill). We refer to this absence as: "The fundamental problem of causal inference." This term precisely means that it is not possible to observe the same entity (e.g., a country, a lab rat, a person, or a household) at the exact same time, with and without the treatment (e.g., with a pill versus without).

This might be frustrating and even discouraging for many people. In my opinion, this is another reason why causal inference is so exciting. The answers are not trivial. Answering causal questions requires technical knowledge but also domain knowledge. The technical knowledge contained in this book will guide your reflection and provide the tools to question causal claims. On the other hand, domain knowledge – your knowledge of the world – will be equally useful to question causality. Hence, when you want to answer a causal question, you always need to dive deep into the topic, try to understand how things are connected, how they work, draw the causal graph, and question your assumptions.

> *Accepting the fundamental problem of causal inference is not the end; it's the beginning.*

This is where it starts being interesting. Even though the ideal situation is impossible to reach, the pursuit of a causal answer involves getting as close as possible to this ideal. This is why I like to refer to the concept of a parallel world as the "key." This is a concept you can always refer to. Then, when you compare this ideal with the situation at hand, it allows spotting the differences and is a useful approach to identify potential issues.

To illustrate how to use this "key," let me use some examples from medical research. Medical research experimenting on animals is one of the setups that allows us to get as close as possible to a parallel world situation, although it might raise ethical questions. Medical research uses rats or mice extensively to study the effects of drugs. But not any rodents – they are usually what we call inbred strains. These animals are bred by mating siblings over more than 20 generations (Encyclopedia of Genetics, 2001 `https://www.sciencedirect.com/topics/biochemistry-genetics-and-molecular-biology/inbred-strain`). In addition to their genetic similarity, their lives are almost identical: they are born and live in a lab, in a cage, with the same food, experience the same types of interactions, and so on. Hence, if we give a drug to one group of those mice and then give a placebo to the other group, we could expect that any difference with respect to a disease is due to the drug because everything else is almost identical. This is known as a parallel group design.

But is it really always perfect? Did we come up with a satisfactory setup for the concept of parallel worlds? Let us use our key to question this example. Remember, we do not actually have access to parallel worlds. What are the other differences in this experiment with mice compared to one with actual parallel worlds? First, we have seen that the animals are quasi-identical. However, imagine that the mice that received the placebo were in a place with more disturbance (next to a door). So, the animals are more stressed, which eventually will affect their behavior and health. We could conceive of numerous other differences that might systematically affect the two groups. Try to think about some differences. It could be with respect to the food (e.g., they forgot to feed the placebo group) or the housing conditions (e.g., the cage was under the air conditioning vent and one of the groups was freezing). This approach is satisfactory only if done perfectly, which is not trivial and will be the subject of *Chapter 6, Randomized Experiment*.

We can even go beyond this example to get closer to parallel worlds in this field of research with inbred animals. Sometimes, it is possible to administer treatment to the animals one day, and then another day give the same animals a different treatment or a placebo. This method is referred to as a crossover design. This method is also used in industry (e.g., Uber) and is often called switchback experiments. See Bojinov, Simchi-Levi, and Zhao, *Design and Analysis of Switchback Experiments* (link: `https://pubsonline.informs.org/doi/10.1287/mnsc.2022.4583`). However, we could still imagine some systematic differences between the moments in which we give the placebo and the treatment. For example, the placebo was given by one experimenter while the treatment was given by another. Imagine that, on the one hand, the team is very cautious about how the treatment, an injection with a syringe in this example, is given because they care a lot about their new treatment. On the other hand, they ask the new intern to do the placebo injection because they think it is less delicate than their newly developed treatment. What if the intern is unsupervised, completely stressed out, and inexperienced, leading to mistreatment of the animals, which eventually affects their overall state? The mistreatment of the subject would negatively affect the mice receiving the placebo, and hence, one might overestimate the effect of the treatment without taking this factor into account. Finally, in some rare situations, we can go even closer to parallel worlds with what we call "within-group comparisons." In this situation, you apply the treatment and the placebo to the same animals at the same time but in different places (e.g., eyes, paws, skin). This almost perfect situation is very rarely an option, as there are many constraints. The main limitation is that we are required to have similar limbs or parts, and treating one part should not affect the others.

The following figure illustrates the different types of designs mentioned:

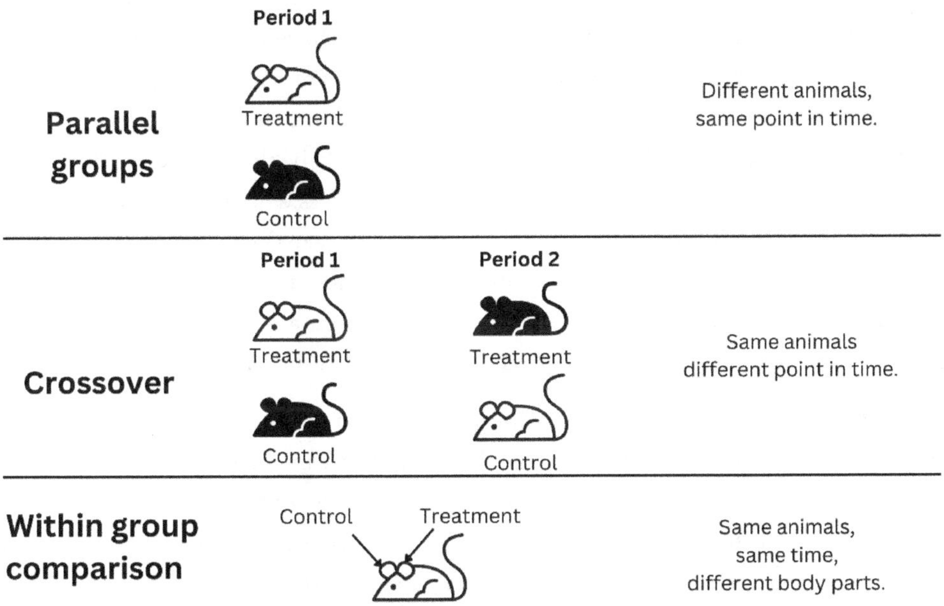

Figure 3.2: Representation of different experimental designs for medical research. The different colors of the mice represent different mouse groups used

You now have a first glimpse of how you can use this "key": comparing the parallel world ideal situation with your situation allows you to identify differences and spot potential issues. But, for the moment, this key looks more like a stick of dynamite used to blow up a safe door. Indeed, we are able to open the safe, but the content is all over the place, meaning that we can spot potential issues, but it is not very clear whether they are actual issues, nor the severity or potential solutions. Throughout the next chapters, we will refine our technique.

The core consequence of this fundamental problem boils down to a single, critical choice we are forced to make. Since we can never observe a truly "parallel world," we must select a real-world substitute for it. This observed situation will act as a substitute for the real counterfactual (the parallel world without the treatment), which we can never observe. I will often refer to this as the proxy counterfactual or control group. This choice is central – it will affect the strength of the claim and the conclusion we make. To do this well, we will continue to dive into this theme in this chapter.

Exploring counterfactuals

Let's revisit this core concept with some more complex, real-world scenarios and, in particular, see that several options are often possible, with pros and cons. The essence of causal thinking relies on a comparison between what happened and what would have happened without a specific element – for example, an intervention, a pill, or a marketing campaign. Hence, to infer cause and effect, you need to compare the situation with the intervention to another reference situation without it. As we have seen, ideally, this other situation should be the parallel world: the treated units, without the treatment during the treatment period. This is what we call the counterfactual. For example, what would happen to my headache if I didn't take the pill? As you now know, we will never observe the true counterfactual (the parallel world). Therefore, to infer causality, we must choose an observable situation to act as our counterfactual. The aim is to find a situation as close as possible to the parallel world. This choice of counterfactual is key.

In the example of the research with mice, what we can use to act as a counterfactual is clear: it is the group of almost identical mice that receive the placebo. In this scenario, we are able to go very close to the parallel world, making the choice of what we can use to act as the counterfactual (proxy counterfactual) straightforward. However, in other instances, the decision can be more complex and multi-dimensional. Currently, there is a lot of debate on the impact of using generative AI (henceforth GenAI) for mental health support. A Harvard Business Review article reported in April 2025 that the number one usage of GenAI in 2025 was therapy and companionship (see: *Marc Zao-Sanders (2025), "How People Are Really Using Gen AI in 2025"*, url: `https://hbr.org/2025/04/how-people-are-really-using-gen-ai-in-2025`). Imagine that you have a chatbot made to help tackle mental health issues. What could you use for the counterfactual to assess the impact on mental health of this chatbot? Is it a group of people not doing anything? Following a standard therapy? Looking for support within their family or on the web? All these different options are valid, but they depend on the questions you are asking, and they will affect the conclusions.

If you want to compare GenAI therapy versus traditional therapy, you should compare people following the two types of therapy. However, if you want to study whether using GenAI can fill a gap by providing support when other options are not available – for example, when you must wait before having an appointment with a busy therapist – in this case, the counterfactual could be "doing nothing." In some sense, this is what was done in a popular study done in 2025. Of the 210 participants, half had access to TheraBot, a GenAI therapy chatbot, while the other half was on a wait list to use the tool for four weeks (see: *Heinz et al., "Randomized Trial of a Generative AI Chatbot for Mental Health Treatment."*, link here, `https://ai.nejm.org/doi/full/10.1056/AIoa2400802`). Note that this study found that people using TheraBot experienced reductions in symptoms of major depressive disorder, generalized anxiety disorder, and clinically high risk of feeding and eating disorders compared to the control group.

Counterfactual thinking is used in various situations. In law, for example, to define a sentence, it is required to estimate the consequence of an action and hence to compare what happened compared to what would have happened without the action. In healthcare, health professionals rely on counterfactual thinking to evaluate the cost-benefit of taking a treatment. In the business world, understanding the effectiveness of a marketing campaign or a strategic decision also requires counterfactual reasoning. The concepts of this book allow us to conduct these exercises properly, reducing the risk of costly errors.

Failing to find a good observed situation to act as the counterfactual prevents us from measuring a causal effect and can thus be misleading. Here is an example I have heard about a company providing free electric scooters in a city. The company claimed it was positively impacting the environment. This claim was based on comparing the emissions from driving a car with those from using an electric scooter and then deducing a reduction in emissions because e-scooters emit less. Hence, the company used car travel as a counterfactual, assuming that free e-scooter services are a substitute for car travel within a city center. Was this assumption sound? Well, probably not: it seems, according to a research paper, that e-scooter services act rather as substitutes for public transport or walking (see: *Reck, et al. (2022), "Mode Choice, Substitution Patterns and Environmental Impacts of Shared and Personal Micro-Mobility."*, url: `https://www.sciencedirect.com/science/article/pii/S1361920921004296`). The conclusions are completely different depending on the counterfactual. If, without the existence of e-scooter services, people would rather walk or use public transport, which generates fewer emissions than e-scooters, then this service actually impacts the environment negatively.

Here is another example: Evaluating the impact of a customer loyalty program highlights the counterfactual challenge: we must find a reliable proxy counterfactual – a comparable group of customers who didn't join – to eventually know what would have happened otherwise. Imagine a store introduces a loyalty card, and after a year, it observes that customers with the card purchase significantly more than those without it. It is tempting to jump to the conclusion that it's working. However, to answer this question properly, we have to understand what the spending of these loyalty card members would have been without the card.

The problem is that the two groups, members and non-members, are not from parallel worlds. They are different customers with potentially different characteristics. Customers who are already frequent shoppers are the most likely to sign up for a loyalty card in the first place. This is a classic case of selection bias, where the "treated" group (card members) is fundamentally different from the "untreated" group (non-members) even before the program begins.

Another comparison could be to compare the purchasing habits of customers in the months before they joined the program to their habits in the months after. But again, using the concept of parallel worlds and the fundamental problem of causal inference, you know that things change over time. In particular, maybe the loyalty program started during the holiday season, when sales are usually higher in your shops. Hence, the before-and-after comparison is also flawed. As this example shows, failing to choose the right control group to act as the counterfactual can lead a business to dramatically misunderstand the success or failure of a marketing effort. Solving this situation is not trivial and would require you to first understand in more depth the issues and the solutions, which is what we are going to cover in the next chapters. If you want to go further, there a scientific paper studying exactly this question. Note that the paper is highly technical. See: *Huber (2024), Causal Machine Learning in Marketing* (url: `https://ijbms.net/assets/files/1721058154.pdf`).

So far, we know that causal claims rely on a comparison: what we observe versus what would have happened without a certain action. Unfortunately, due to the fundamental problem of causal inference, the absence of parallel worlds, we do not have the ideal counterfactual.

Hence, we must find an often imperfect observed situation to act as the counterfactual. We have just seen with varied examples how identifying the counterfactual and comparing it to the ideal parallel world situation allows us to detect some limitations or problems. This is already an incredibly powerful tool you can use, and it will be the first step in evaluating causal claims.

So far, I have explained causality and how to challenge it in abstract terms. The approach I use leans more towards critical thinking and thought experiments than towards statistics, even though my formal education focused on the latter. There are two reasons for this. First, I have realized that a big part of what I am doing does not require a single equation. This approach allows me to share the essence of these concepts with a broad audience and to popularize this knowledge. Second, you can't directly test the presence or absence of causality in data without making any assumptions. This is a direct consequence of the fundamental problem of causal inference. Since there is no way to access a parallel world, statistical tests cannot resolve this issue. In order to have a causal effect, the most important assumption, called the unconfoundedness assumption, is usually not testable. The unconfoundedness assumption means the following: the group receiving the treatment (e.g., a coupon) and the group not receiving the treatment would be in the same state (e.g., the same level of spending) if neither group received the treatment. If this assumption is true, then we have a proper way to identify a causal effect. However, since we cannot observe the treated group without the treatment during the treatment period, we cannot directly verify the validity of this assumption. We can only assess how plausible the assumption is – for example, by checking that the people in the two groups do not have systematic differences.

If we cannot prove direct causality with a test, what can we do? Let's answer this question now.

Proving causality

Now we have reached the bottom of the valley of despair. We have seen that causality is essential, but our enthusiasm fell like an ice cream cone on a hot summer day held by a tired three-year-old at the swimming pool. Luckily, we will now start to climb back up the knowledge curve. We will rise from this valley by climbing numerous steps. The following example will reveal the general idea of how we can find evidence of a causal effect before going into more detail in the rest of the book.

Although we cannot prove causality with a simple test and verify the unconfoundedness assumption directly, we can actually do many things to test the credibility of this assumption. To illustrate this important aspect, we will see how researchers proceed using as an example a fascinating scientific paper studying the effect of air pollution on mortality in London (see *Hanlon (2024), London Fog: A Century of Pollution and Mortality, 1866–1965.*, url: `https://direct.mit.edu/rest/article-abstract/106/4/910/111191/London-Fog-A-Century-of-Pollution-and-Mortality`).

London was a densely inhabited and polluted area already during the 19th century. The author, Walker Hanlon, answers a critical question: What is the effect of pollution exposure on mortality? However, here comes the first issue: air pollution data has only been available since the 1950s. To address this challenge, Hanlon came up with a creative idea, taking advantage of the accurate meteorological data available since the 1850s. His idea was to use fog as a proxy for pollution because when London is under heavy fog, pollution remains at low altitude, and citizens' exposure to pollution increases.

Now, we will put ourselves in the shoes of police detectives. Police detectives continually investigate causal questions, including who is responsible for a person's death. Was it Colonel Mustard with the candlestick in the conservatory? Are you sure that the crime was not committed using a wrench or by someone else? In our case, we have a suspect, or more precisely, a hypothesis that we would like to test: fog increases mortality due to pollution retention. Then we ask ourselves: Was it really pollution or the evolution of health services? Or, was it actually a direct consequence of the weather? In this situation, all these factors affect mortality. But we want to pin down precisely the role of one suspect: air pollution. More specifically, we will focus on one part of the paper showing that heavy fog increases mortality in London due to the increased pollution concentration.

Video footage of someone committing a crime is rare, making it challenging to identify the perpetrator. Even if you do have such material, maybe the image is blurry, or maybe it's a fake. Therefore, you might never be one hundred percent sure of the identity of the criminal. To overcome this limitation, you accumulate evidence, and you try to discard all the possible alibis of the criminal until you have enough evidence. In research, when we want to find causal evidence, we employ the same approach. We explore alternative theories explaining the effect we observe and try to discard all those stories until there is only our hypothesis left.

The main aspect of the publication is the investigation of the effect of heavy fog on mortality rates. Simply looking at how the mortality rate (all causes together) changes during and after a week with heavy fog is problematic, as numerous other things are different. Remember the discussion about the choice of the control group or proxy counterfactual. Many things might change before and after, when there is heavy fog and when there is not. In particular, we can see from the *Figure 3.3* left-hand-side graph that, on average, there were fewer weeks of heavy fog during the 20th century, while the *Figure 3.3* right-hand-side graph reveals that fog is almost nonexistent during spring and summer (between weeks 14 and 39). If there is only fog in fall and winter, the excess deaths could be due to the cold and not pollution.

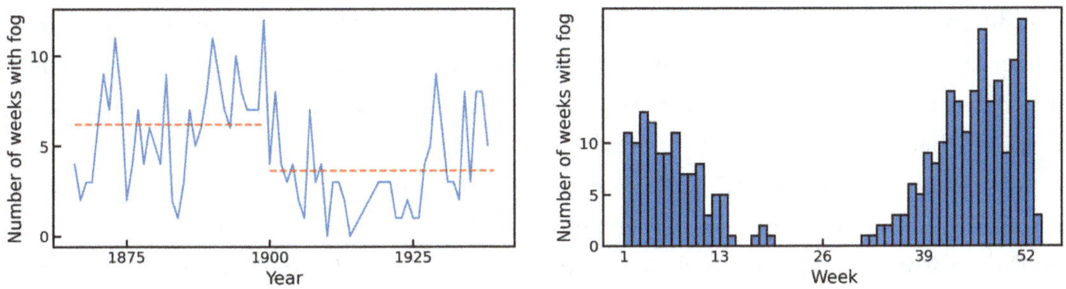

Figure 3.3: The first graph represents the number of weeks with heavy fog by year. The horizontal red dashed line represents the average number of weeks with heavy fog per year before and after 1900. The second represents the number of weeks with heavy fog by calendar weeks

Note that Hanlon demonstrates that, while smoke particles can, in theory, promote condensation, the timing of London's dense fog was overwhelmingly dictated by specific atmospheric conditions. Short-run variations in coal emissions did not predict when the fogs formed, so the reverse effect of fog on pollution can be considered marginal for his analysis. This is a common problem called reverse causation, which will be discussed in the next chapter.

Armed with our causal inference lens, we see that pollution, fog, and respiratory disease are connected in an intricate web of causal relationships. To disentangle this and identify whether pollution is indeed the culprit, increasing mortality, we need to play the role of the detective and challenge the alibi of our suspect.

Playing the causal detective game

Luckily, it is possible to take into account those factors in a statistical model, allowing us to disentangle the effect of heavy fog from the effect of seasonality and time trend. For example, in order to account for the evolution between centuries, instead of comparing the absolute level of mortality over time, we will look at the deviation from the average weekly death rate each year. A deviation means the distance from the average. This approach allows us to compare things that are comparable. Mortality in 1870 might be different from 1935 due to the quality of the healthcare system, for example. Hence, only comparing the death rate deviation from the weekly mean within the year 1935 and the deviation from the weekly mean within the year 1870 takes into account the general context of each year and prevents confounding the effect of fog with the effect of the evolution of healthcare. In essence, we are changing the proxy counterfactual here. Instead of comparing between years, we study the within-year variation.

In addition, we can "control" for other factors such as seasonality or weather conditions (e.g., temperature). "Controlling" means that we keep it constant, allowing us to estimate only the effect of a change in our main cause of interest. In other words, we can include in a statistical model a variable that will capture the effect of seasonality, as well as the effect of weather conditions such as temperature, on mortality. This will allow us to isolate their effects and prevent them from being mixed with the effect of fog in our model. To go beyond this intuitive explanation requires relatively advanced knowledge of statistics. I acknowledge that this is a limitation of the approach used in this book. Nonetheless, you will see that you can grasp all the concepts in this book without this technicality, and without years of studying statistics. However, if you do want to open this box and dive into the technical aspect, I recommend two books: *Causal inference: The Mixtape*, by Scott Cunningham, Yale University Press, and *Causal Inference in Python: Applying Causal Inference in the Tech Industry*, O'Reilly, by Matheus Facure. The first is more academic, while the latter is more oriented toward the business world.

Figure 3.4 shows the first result: mortality increases during weeks with heavy fog and in the week after. The vertical axis represents the death rate (all types of deaths together) while the horizontal axis represents the distance in weeks from a week with heavy fog. The dots represent the average deviation of the number of deaths within a year, as we are looking at the deviation from the mean (excess mortality) to compare what is comparable (death within a year and not comparing 1925 with 1875, where the overall death rate was probably higher in 1875 due to sanitary conditions and fog as well, as we have seen earlier). For example, during week 0 (a week with heavy fog), the number of deaths increased by about 70 deaths on average. The vertical bars above and below each point represent error bars. Simply put, if these bars cross the value 0 on the *y* axis, it indicates that the variation is not statistically significant—in other words, the difference could be due to random fluctuation. Hence, we can see that only the weeks with a statistically significant difference are the weeks with heavy fog, and the one after. If this is a story about health issues due to exposure to pollution, it means that people would have a higher chance of dying during a week with heavy fog and possibly up to one week later.

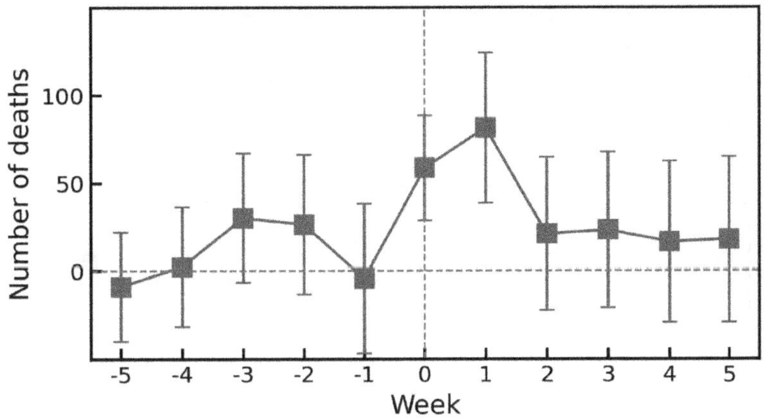

Figure 3.4: This figure reports the number of deaths per week around a week with heavy fog, once the seasonality, weather, and time effects have already been captured by the model

But remember, we should question everything and act as if we were police detectives. Is it really about health? Could it be that during weeks with heavy fog, there are more accidents on the road due to bad visibility, or more crime as people can hide more easily?

To rule out these alternative stories, the author compares the number of deaths based on the recorded cause of death (e.g., accident/crime versus pneumonia). In *Figure 3.5*, the squares represent the deaths of people who died due to an accident or crime, while the circles represent the deaths of people who died from pneumonia. Note that pneumonia is a lung disease that can be triggered or aggravated by pollution, which is why we make this comparison.

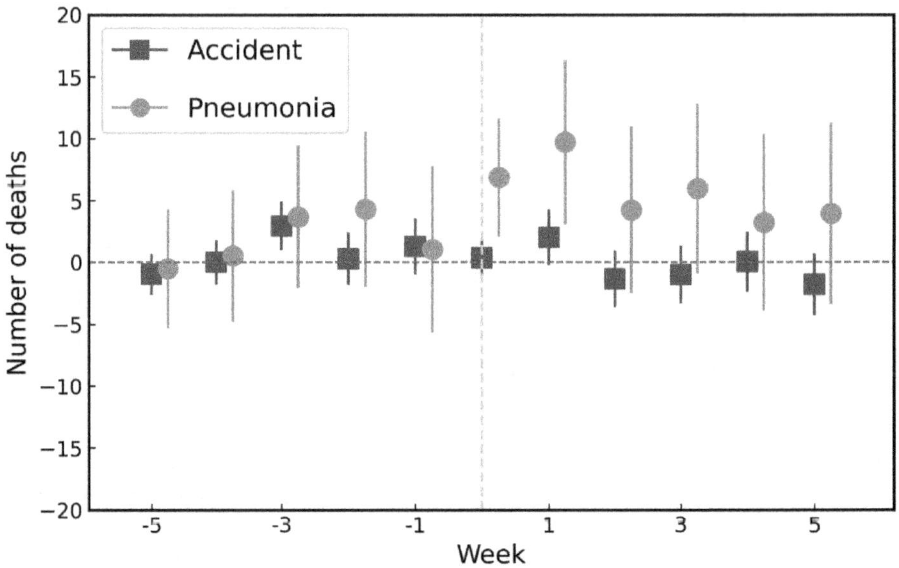

Figure 3.5: This figure reports the additional mortality from accidents/crime versus pneumonia around a heavy fog week

Figure 3.5 reveals that there is no statistically significant change in the death rate caused by accident or crime around a week with heavy fog. This confirms that the main driver is respiratory disease, arguably triggered by the increase in pollution. We are getting closer to catching our suspect: pollution. There is one last main alternative story we might want to discard: Could the increase in mortality instead be caused by the fact that people stay at home and contaminate each other when there is heavy fog? When the weather is bad (foggy), people might stay home. Hence, if people stay at home, there is a higher risk that they will contaminate other people with viral respiratory diseases and therefore the increase in deaths is simply a consequence of that and not of pollution. It seems very difficult to falsify this story, right?

The author of the paper, Walker Hanlon, did something very elegant to dismiss this alternative story. He compares two different weather shocks: heavy fog and heavy rain. Indeed, heavy rain is likely to have a similar effect on people's behavior: people might stay at home. However, the key point here is that fog traps the pollution while rain cleans the air. Therefore, if this is a pollution story, we would find the opposite effect of rain (fewer deaths) compared to fog. The following figure clearly shows this effect: while fog increased mortality, rain appears to have had a lifesaving effect in 19th-century London.

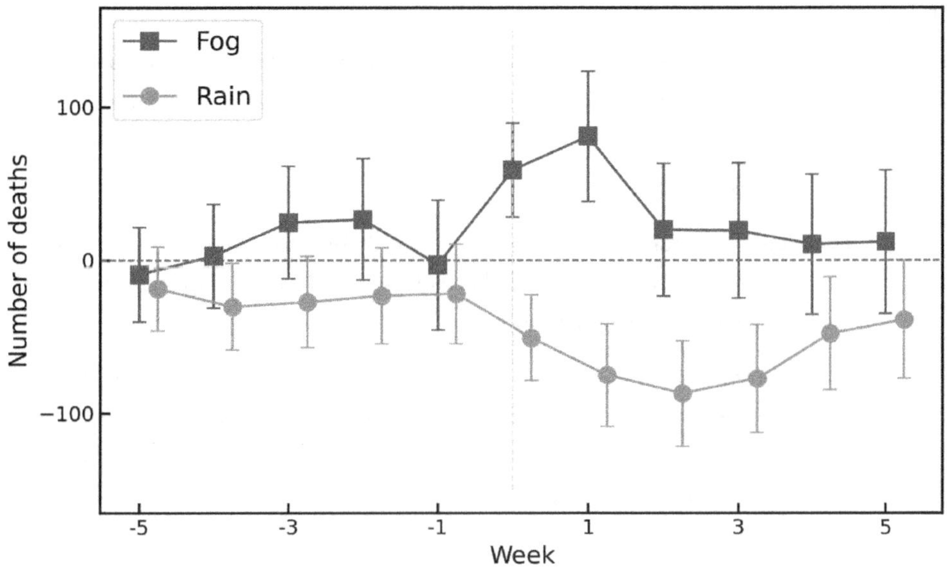

Figure 3.6: This figure represents the overall mortality rate around a week with heavy fog (squares) compared to the mortality rate around a week with heavy rainfall (dots)

We have been able to discard several alibis (e.g., bad weather forces people to stay at home and hence might increase the spread of diseases) and other suspects (e.g., weather, evolution over time, accidents, or crimes). As explained in the previous chapter, explaining the scientific process, imagine that this paper has been challenged by experts numerous times, potentially for years, where they tried to find all the potential alternative explanations, and the author had to answer those questions. Ultimately, the author piled up enough evidence to say that the effect we observe is indeed caused by pollution.

This paper illustrates how to address the fundamental problem of causal inference in complex settings and the fact that you can't prove direct causality with a statistical test. Despite this absence, we were able to witness that, indeed, data and statistics are useful in proving our claim.

For now, you already own your key to open the world of causal inference: the parallel world story. This key can be used to identify the difference between the proxy counterfactual you are using and the ideal parallel world. Remember, the ideal setup would be to have parallel worlds in order to have only one difference between the two worlds (the cause), everything else being identical. Hence, if we had parallel worlds with just one difference and we observe a difference with respect to the outcome, it would be possible to claim that it was due to the cause, as it was isolated. But the world is complex, and hence, as a police detective, we have to discard or isolate all the other potential explanations in order to see whether our suspected cause survives. This is the broad idea.

In the following chapters, we will see how to go through this process as efficiently as possible. The next chapter will provide the toolkit of a causal forensic expert. We are going to arm ourselves with a set of questions, allowing us to challenge causal statements in a structured way.

Key concepts

The "key," the parallel world story: The only way to perfectly measure a causal effect would be to have two parallel worlds, and we have the capacity to change just one thing in one of the two worlds.

The fundamental problem of causal inference: We do not have access to parallel worlds for experimentation. Hence, we are unable to observe the same entity (e.g., country, person, animal) at the same point in time with and without the intervention (e.g., weapons import, doing regular physical activity, taking a pill).

Counterfactual: The counterfactual is the alternative world without the intervention.

Proxy counterfactual: The proxy counterfactual is the observed situation we use to act as a counterfactual. It is also often referred to as the control group.

Unconfoundedness: In order to measure a causal effect, we must isolate the cause from other effects that could be confounded. Hence, we can assume that the group receiving the treatment (e.g., a coupon) and the group not receiving the treatment would be in the same state (e.g., the same level of spending) if neither group received the treatment.

Case studies

For the three case studies, think about whether a valid control group or proxy counterfactual could be used to estimate the effect. Using the parallel world story, try to identify the potential differences between the parallel world and your counterfactual. For example, if you want to assess the effect on sales after an ad campaign done in summer, and you look at sales in winter when you didn't have an ad campaign, the main difference there is the season.

Next, I present some potential answers. You might come up with different counterfactuals and points. There are countless options. As we advance in this book with the case studies, we will eventually narrow this down, aiming for the ideal setup but also something feasible.

Case study 1: Cold showers and the immune system

What if we compare the health of people who take cold showers with those who do not take cold showers?

People who take cold showers might be, on average, more invested in a healthy lifestyle for two different reasons. First, the fact that someone knows about the potential benefits of cold showers might reflect their interest and education level on the matter. Second, if someone is willing to do something uncomfortable for their health, it signals that they are invested and potentially would take other measures to stay fit and healthy. Hence, this systematic difference creates a selection bias and prevents assessing causality.

Case study 2: Marketing campaign for energy sobriety

In 2022, to answer the threat of an energy shortage in Switzerland, a large governmental campaign to promote energy sobriety was launched. The campaign consisted mostly of a vast advertising campaign with energy-saving tips and recommendations. A graph follows, representing the monthly total energy consumed in Switzerland the same year. The vertical dashed line represents the beginning of the marketing campaign for energy sobriety.

Using the month before as the proxy counterfactual for the consumption is problematic. The pattern of consumption is subject to seasonality. The temperature varies between months, and in Switzerland, an important part of energy consumption in winter is for heating. Hence, the increased consumption during winter is just a reflection of the change in weather conditions and will not allow us to measure the impact of the campaign.

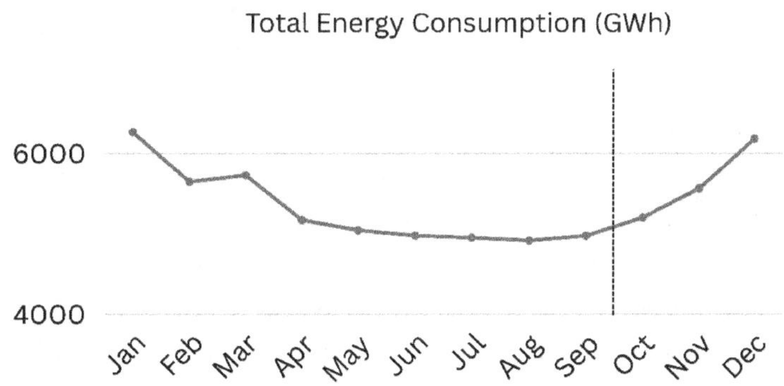

Figure 3.7: This figure represents the monthly total energy consumption in GWh in Switzerland in 2022

Case study 3: Air pollution and respiratory disease

Imagine now that to assess the effect of air pollution on health, we compare the prevalence of respiratory disease between heavily polluted areas and less polluted areas.

There might be other differences between those areas. People living in heavily polluted areas could have a lower socio-economic status on average. Hence, they might differ in their education level, income, or even access to health services.

Another comparison could be between urban and rural areas, assuming that urban areas tend to have more air pollution than rural areas. However, comparing the risk of respiratory disease, known to be aggravated by air pollution, between rural and urban areas is not adequate, as rural areas are not a good counterfactual. People living in rural areas might systematically differ from people living in cities with respect to their demographics, but also their daily lives (e.g., job, diet, physical activities). Hence, it would not be possible to isolate air pollution as the unique factor since other variables also affect the relationship.

Summary

This chapter took us into the "valley of despair" to reveal the fundamental problem of causal inference. You learned that because we can never observe a person, a business, or a lab rat in two states at once, with and without an intervention, we miss the ideal setup to measure causal effects. To overcome this, we must find a real-world substitute for the ideal but impossible "parallel world," a concept we call the proxy counterfactual or control group. We saw through several examples, from e-scooters to customer loyalty programs, that the choice of this proxy counterfactual is a critical decision that can dramatically challenge the quality of the conclusion or simply reframe the answer.

This information is crucial because it transforms you from a passive consumer of information into an active detective. Armed with the "parallel world" key, you can now begin to question any causal claim by asking, "What is being compared here, and how different is that comparison from the ideal?" While we learned that we cannot directly prove causality—in particular, the unconfoundedness assumption—with a simple statistical test, the detailed case study of the London fog showed how we can collect robust evidence by systematically ruling out alternative explanations, getting us closer and closer to the true cause.

Now that you understand the core problem and the detective-like mindset needed to approach it, we are ready to climb out of the valley. The directed graphs at the end of *Chapter 1* could help us notice differences. In the next chapter, we will equip you with additional tools. We will pin down exactly which differences between your observation and your proxy counterfactual are the most problematic, introducing you to the source of flawed causal claims: confounders, reverse causation, measurement errors, and threats to external validity.

4

Beyond Correlation: The Main Culprits

The last section highlighted why it is so hard to assess causality due to the fundamental problem of causal inference and laid the foundation for identifying the issue. Do not despair – we will now start our ascension of the slope of enlightenment by expanding our toolkit, allowing us to identify the potential issues in this chapter and exploring solutions to those problems in the next.

In this chapter, you will learn three questions you can always ask to challenge a causal claim. We will see that the first two allow us to identify precisely why a correlation might be misleading. The third question will be slightly different, but equally important, as it will connect to one of the most common mistakes: misleading extrapolation. Sometimes, people will find a clean causal effect, but their conclusion will be wrong. This is especially true when people rely on scientific research and use it as an authoritative argument. We will see precisely how to detect this issue and invalidate these claims. Hence, in this chapter, you will enrich your toolkit to challenge causal claims with three new powerful tools. Combined with the concept of a counterfactual and directed graph, it will complete your causal forensic toolkit used throughout this book.

In this chapter, we will cover the following topics:

- Identifying confounders
- Spotting reverse causation
- Questioning external validity

Identifying confounders

Let me start by giving you the most important and powerful tool I use to question the validity of causal claims. It might sound so simple at first, yet it is profound, full of subtleties that we are going to explore. You will be able to sharpen it with practice to go deeper. This tool, as you may have guessed from the previous chapter, helps us question the unconfoundedness assumption—the idea that the effect of interest is not mixed with something else. It builds on a key insight from the last chapter: because we do not have parallel worlds, there are almost always systematic differences between treated and untreated groups.

The first question that you can use as an additional tool to challenge causal claims is the following: "Is there something else explaining or affecting this relationship?". This question asks whether there might be another factor confounding the observed effect. Remember, in *Chapter 1*, we saw that the main reason why correlation does not imply causation is that you have a third variable affecting both your cause and outcome. This is exactly what I am referring to. This third variable provides an alternative explanation for the observed correlation. Shark attacks do not cause ice cream sales; both increase when the weather is sunnier. Similarly, ads might increase sales, but the observed effect could be inflated by seasonality. We actually already used this tool in the last chapter, when studying the effect of fog on mortality in London. We explored whether there was "something else" explaining the increase in deaths during weeks with heavy fog. We asked whether it could be explained by the fact that it is usually colder during the period with heavy fog.

We have an issue if there is something else that we do not take into account affecting both the potential cause and the result we are interested in. Technically speaking, this is called an **omitted variable bias**, as it biases the relationship between the cause and consequence if we omit to take it into account. This problem is also often referred to as a *backdoor path*. Using our directed graphs, we have to be worried only when these two conditions are met: 1) this other variable (W) affects the cause (X), and 2) the outcome (Y). Here is a representation with a directed graph:

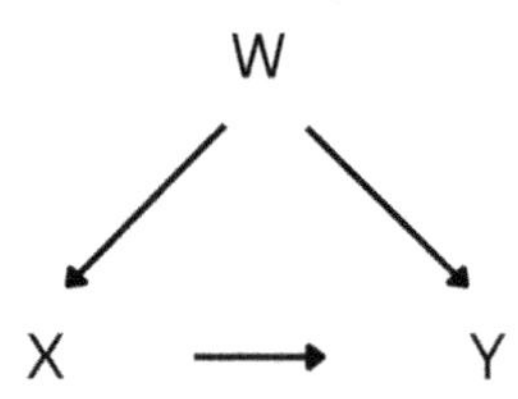

Figure 4.1: Representation of a confounder W, affecting the relationship between X (the cause) and Y (the outcome)

Remember the examples we saw when discussing causal directed graphs in *Chapter 1*. As illustrated by the preceding graph, the fact that ice cream sales (referred to as **X** in *Figure 4.1*) could be positively correlated with shark attacks (noted as **Y** in the figure) due to the confounding effect of weather (represented by **W**): when it is sunnier, people consume more ice cream (condition 1) and also go swimming, increasing the risk of potential incidents with sharks (condition 2).

By contrast, if you have another factor that affects either your cause of interest or your outcome, but not both, it will not challenge your causal claim. Let me illustrate this with the case of a food stand at a music festival. Imagine you hire a new digital marketing expert. This expert's skill will affect your ad spending and ad quality (the cause), but it will not directly affect your sales (the outcome). The expert's impact on sales will only happen through the marketing campaign. Alternatively, consider a sudden, unpredictable shock that affects only your sales. For example, if it starts pouring with rain, your sales will likely go down. But because this is a short, random shock, you won't necessarily change your social media marketing strategy in response. Therefore, the rain affects only your outcome (sales).

In both of these situations, the causal relationship between your marketing and sales is not confounded. Remember, a confounder must meet two conditions: it must affect the cause, and it must also affect the outcome. Here, neither the marketing expert nor the rain meets both conditions, so they are not confounders. This is depicted in the following figure:

Figure 4.2: Directed graph representing a situation without confounders

Nobel prize laureates and chocolate consumption

A very popular example of spurious correlation is the strong positive association between chocolate consumption and the number of Nobel prize laureates, as reported in Messerli (2012) *Chocolate Consumption, Cognitive Function, and Nobel Laureates* (link: `https://www.nejm.org/doi/abs/10.1056/NEJMon1211064`). This document led to numerous other research papers. Here is an approximate replication of the main graph in the article:

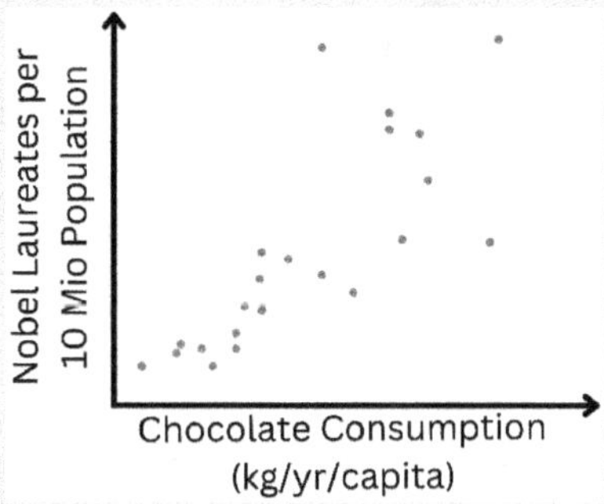

Figure 4.3: Approximate replication of the correlation between chocolate consumption and Nobel laureates, as depicted in Messerli (2012) "Chocolate Consumption, Cognitive Function, and Nobel Laureates." Each point represents a country

The initial "occasional note" by Messerli in *The New England Journal of Medicine*, one of the most respected medical research journals in the world, stated this *"Of course, a correlation between X and Y does not prove causation but indicates that either X influences Y, Y influences X, or X and Y are influenced by a common underlying mechanism."*. However, the author continues right after this sentence with a suggestion of a potential causal pathway: Cocoa contains flavanols, chemical compounds found in varied aliments, which are known to positively impact the brain and cognitive function. While this is tempting to believe to justify our chocolate consumption, this causal claim has been debunked.

One year later, a paper (Maurage et Pesenti, 2013) was published to challenge the claim: *Does chocolate consumption really boost Nobel Award chances? The peril of over-interpreting correlations in health studies* (link: `https://www.sciencedirect.com/science/article/pii/S0022316622012196`). In this paper, they replicate the same correlational analysis with other flavanol-rich foods and find no relationship, and in addition, show the very strong correlation between chocolate consumption and GDP, suggesting that it is potentially the main confounder or source for this spurious correlation.

A related central issue in causal inference that can be seen through the same lens is that of selection bias. Selection bias arises when the groups receiving a treatment have systematic differences that also affect the outcome. A common example used in economics for this is the one relating class size in schools and academic performance. Having smaller classes (fewer students per teacher) allows for more individual interactions with the teachers. This more efficient learning environment could eventually increase the academic performance of the students. However, imagine that, to study this relationship, you compare the results from students from small classes against those in bigger classes and find that the academic performance for the students from smaller classes is better on average. Fantastic – it means that, indeed, having a small class is beneficial! Or is it? Now, you are becoming cautious with such claims, right? In particular, you can use your first tool: "Is there something else explaining or affecting this relationship?".

Socio-economic factors could be confounded with the effect of the class size on the results. If population density affects class size and is related to socio-economic factors, it could affect the effect measured. Imagine that people with lower income, live in places more densely populated if the rent is lower. The parents of the students might have fewer opportunities to help the children at home (for example, due to lower education levels and a lack of the necessary knowledge, or because they are too busy juggling two jobs). In addition, lower income might be synonymous with fewer rooms in the house and fewer technological resources. Hence, the conditions to study at home might be far from ideal (e.g., one computer shared with siblings, no room to study quietly).

In this situation, the negative association between class size and academic performance could be due to class size, to socio-economic differences, or both. This systematic difference is a selection bias that implies that the effect of class size is confounded with that of socio-economic status. It is called a selection bias because the allocation between small and larger classes is not random. The wealthier are more likely to be allocated (selected) in a smaller class. See the following figure for the corresponding directed graph:

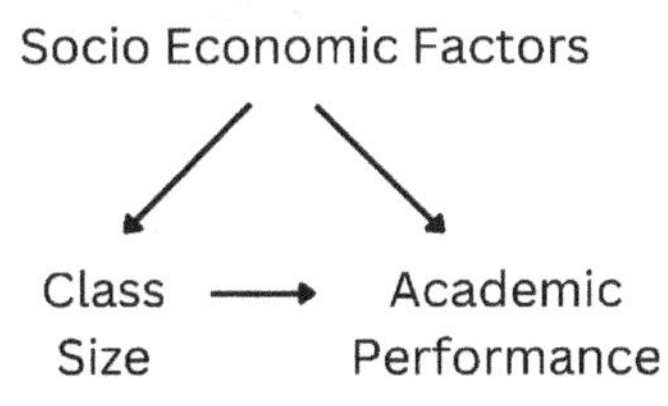

Figure 4.4: Directed graph between class size and academic performance

It is essential to stress again that this problem of confounding effect arises only if the arrows are pointing at your cause of interest and outcome, as we have seen on the last two directed graphs. Cause and outcome can be connected in other ways than this. Let us now illustrate the other potential relationships.

Mediators and colliders

First, let me reuse the example of roadblocks by environmental activists from *Chapter 1* to remind us of another common and important relationship in directed graphs:

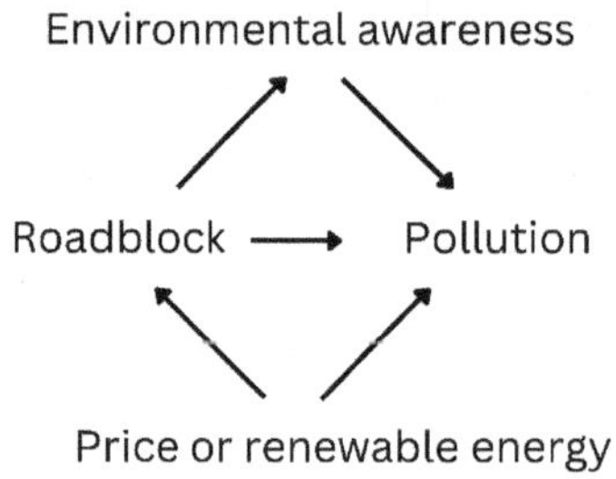

Figure 4.5: Directed graph representing the relationship between roadblocks by environmental activists and pollution

In the example of the roadblock, the price of renewable energy is arguably a confounder. This price has been drastically decreasing in recent years, which might encourage activists to fight against the usage of fossil fuel (first condition), and allow the reduction of pollution (second condition). On the other hand, the arrows connecting **Environmental awareness** with **Pollution** are not both pointing to the cause (**Roadblock**) and the outcome (**Pollution**). Hence, this is not a confounder as the two conditions are not met here! In this case, the arrow goes from the cause (**Roadblock**) to the outcome (**Pollution**), passing by environmental awareness. This is what we call a **mediator**. A mediator does not question the causal relationship; it reflects indirect pathways. A mediator allows us to understand how the cause is affecting the outcome and is depicted as follows:

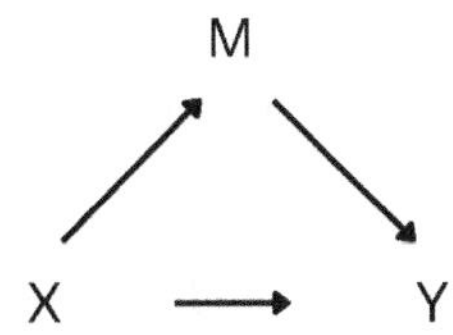

Figure 4.6: Directed graph representing a mediator (M) with respect to the cause X and outcome Y

If both the cause and the outcome point to a third variable, that variable is not a confounder but a **collider**. In this case, the arrows from the cause and the outcome "collide" on a third element. A collider should not be included in a causal model, or it will actually bias this result. For example, diabetes and lung diseases are two (generally speaking) unrelated health issues. Both affect the probability of hospitalization; hence, it is a collider. While these two health issues are unrelated, the inclusion of a collider would create a correlation in a causal model and hence wrongly show an association between the two. For example, if we focus only on hospitalized patients, we might mistakenly see a connection between diabetes and lung disease, even though there is no direct link. This is because if you suffer from both, the chance that you will be hospitalized is even higher, which makes it look like the two conditions are related. While this is a highly technical topic, the main point is that you should not try to take into account or "control" a collider. You can refer to Tönnies et al. (2022), *Collider Bias in Observational Studies* (URL: `https://pmc.ncbi.nlm.nih.gov/articles/PMC9131185/`), for more information. A collider is represented as follows:

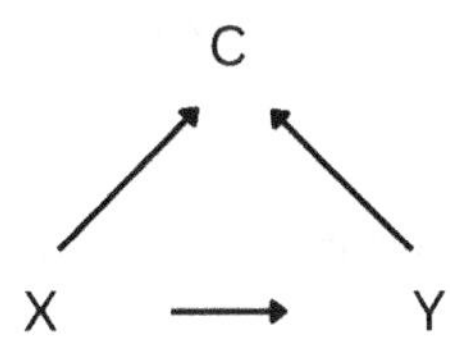

Figure 4.7: Directed graph representing a collider (C) with respect to the cause X and outcome Y

Advanced technical concept: feature selection

As we saw in *Chapter 2*, in predictive inference – for example, with machine learning – the goal is to select features in a model that improve the quality of the prediction. This process can be done in a data-driven way. You can give a large set of potential predictors to an algorithm and let the algorithm choose the best predictors.

This is not possible with causal inference. In causal inference, your model must include the cause, the outcome, and all potential confounders. Including the confounders allows you to isolate their effect, separate them from the impact of your cause of interest, and pin down your causal effect. Hence, the rationale for the inclusion of "features" in your model, in causal inference, relies extensively on domain knowledge and on your causal graph. This can't be done in a fully automated way. Some of the features might be poor predictors, with potentially no statistically significant effects. These types of features are often excluded in predictive inference but not in causal inference. Their direct effect is not what we are interested in. We care about how the inclusion of confounders affects the impact of the cause of interest on the outcome. Moreover, in causal inference, there are some features that you should not include, or, instead of resolving issues, they will create bias. This is the case with colliders. A collider should never be included in a causal model, or it will mess up the estimation. If you want to go further, you can read a famous paper on the topic, *A Crash Course in Good and Bad Controls*, Cinelli et al, 2022 (link: `https://ftp.cs.ucla.edu/pub/stat_ser/r493.pdf`).

Finally, there is a less common and more subtle situation related to confounders that we should cover.

Non-classical measurement errors

Sometimes, what we want to measure is not well measured, which can create estimation problems. This is what we call measurement errors. Simple examples include an imprecise tool, such as a thermometer that sometimes overestimates and at other times underestimates the temperature randomly. This also happens in surveys. If you ask people how often they use your app, they might not recall well or may tend to give more common answers, such as "10 hours" rather than "8.75 hours" per week. If you study crime or illegal markets, you often won't observe all the crimes, but just the ones reported; you won't observe all the criminals, but just the ones who get caught. In mental health, we often don't measure the real incidence of some issues but just the reported, observed incidences.

I remember discussing with a university dean the fact that there was a growing number of students seeking mental health support, and this person thought it was very worrying. However, I also pointed out another possibility: mental health services may simply be more accessible now, and the declining social stigma makes it easier for people to seek help and report doing so. The issue here is that we do not observe the number of psychological issues, just the people seeking help. Moreover, if the quality of detection or the definition of an issue changes over time, it might inflate or reduce these numbers while the underlying problem is stable or even moving in a different direction.

While this is a broad topic, I want to break it down and focus here exclusively on the effect on causal claims. When it comes to causal inference, we classify these issues into two categories: classical measurement errors and non-classical measurement errors.

Classical measurement errors refer to situations where measurements suffer from random imprecision. The first examples I shared – an imprecise instrument over- or under-reporting a measure, or clients in a survey imprecisely estimating time spent on an app – would belong in this category. This type of issue adds, in statistical terms, noise. If the measurement error is in the outcome, this random variability makes the true causal effect harder to detect but does not bias the estimation. If the measurement error is in the cause of interest, it creates an **attenuation bias**: your effect will systematically appear smaller than it really is.

The more serious issue when it comes to causal inference is non-classical measurement error. This issue arises when the magnitude of the measurement errors are systematically related to our cause or outcome. Let me illustrate this situation with a concrete example. During the COVID-19 pandemic, we didn't observe the number of people sick; we only observed the number of people who tested positive. Moreover, this was not just random noise. Countries would change their testing policies and rates depending on the situation. In Switzerland, at some point during the first few months, when the number of cases was very high, we feared a shortage of tests. Hence, the policy was to reduce the testing rate to save tests for when it was most necessary. However, if you reduce the number of people tested, you will also reduce the number of cases detected. This downward bias is not random; it is directly related to the total number of actual cases.

Here is where this example becomes interesting. When the first vaccines appeared, a graph was published in Switzerland showing the number of cases per 1,000 people by region alongside the vaccination rate. This graph revealed a clear negative relationship. However, critics argued that this could be due to testing rates and non-classical measurement errors. At the time, if you were not vaccinated, you needed to get tested to go into many public places. Hence, in places where the vaccination rate was low, more people got tested, potentially allowing for the detection of more COVID-19 cases, which could contribute to the observed negative correlation.

This is a case of non-classical measurement error, as the testing behavior (and thus the measurement accuracy) was systematically linked to the vaccination rate (our cause of interest).

A journal addressed these criticisms in an interesting way. They remade the graph but included a third piece of information: the color of the dots would represent the positivity rate (the percentage of people who tested positive among those tested). This addition revealed that in regions with a high vaccination rate, the positivity rate was very low, while the positivity rate was much higher in regions with a low vaccination rate. This suggests that the relationship observed was not primarily explained by differences in testing. This additional information implies that regions with a high vaccination rate were likely testing enough (indicated by the low positivity rate), while regions with a low vaccination rate were still not testing enough, as the high positivity rate suggested more widespread infection.

I decided to include the theme of measurement error within this broader discussion of confounders for two important reasons. First, in my experience, severe problems arising from measurement error are less common than the other issues we will discuss, namely, confounders, reverse causation, and external validity. Second, and more importantly, we can use the same critical question to detect measurement errors that we use for confounders: "Is there something else affecting or explaining the relationship I observe?" In this case, that "something else" is the measurement error itself, so the core investigative tool works perfectly.

So far, we have seen how a *third element* can complicate or explain a causal claim. However, there is another very important situation that can challenge our conclusions. This next scenario is different as it does not require a third element but instead relates to the direction of the arrows in our causal graph.

Spotting reverse causation

So far, we haven't questioned the direction of the arrow between the main cause of interest and the outcome, but this could clearly influence the causal relationship and conclusion. Imagine seeing a positive correlation between the crime rate and the number of police officers patrolling. Someone could accuse police officers of being the cause of crime. While the relationship between crime and the police force is certainly not simple, we can claim that it is more likely that the relationship works the other way around: because there is more crime, more police officers are patrolling. This is what we call reverse causation.

Reverse causation is a common problem in numerous situations. When I was 10 years old, I saw a documentary on TV about a very energetic 95-year-old man. The reporter asked him what his secret was. The man, very simply, said that he swam every single day in the cold lake in front of his house. As a child, I was so excited. I felt like I understood it all. I knew the secret to life and longevity. So, I rushed upstairs to tell everything I've just learned to my older brother. After I finished explaining, without even looking at me, my brother casually said: *"It's nonsense, it's rather the opposite. The guy has good genes, he is dedicated, and that's why he is able to swim every day"*. Essentially, without realizing it, he was pointing to the issue of reverse causation.

I remember this scene vividly, even though it is now more than two decades ago. Without knowing it, it probably contributed to my path towards my career, passion, and obsession today with causal inference. Or, was it because I had this underlying interest that this event remained stuck in my memory? Could it be the reverse?

Another situation that I will also include in this terminology, for simplicity, is simultaneity. Simultaneity happens when two elements affect each other causally. Let me illustrate this concept with two situations that I encountered during my academic career.

When I started my Ph.D., I tried to answer what I thought was one of the most important questions at that time: *"How does sending weapons increase the risk of conflict in the destination country?"* In recent history, the most prominent type of conflict is internal conflict (conflict within the borders of a single nation), particularly in Africa.

Many people will say that there is no debate; it is obvious that weapons' availability increases the risk of use of force, violent repression, or rebellion. Actually, important reports published by the Red Cross back in the 90s made this point clear and observed a very strong correlation between the presence and intensity of violence and the availability of weapons. However, as you know, correlation does not imply causation, and it is insufficient to report a strong correlation. This problem allows arms lobbyists and exporters to argue that they are not the cause of the problem. We could say that weapons do not cause violence, but when there is a violent situation, people buy weapons. While it is misleading to claim that weapons are merely a consequence of war without influencing the intensity or risk of conflict, it is undeniable that the demand for weapons increases as tensions rise. Using causal inference terminology, this is what we call reverse causation (or simultaneity). Both weapons and wars affect each other, which might create a mutually reinforcing loop, as depicted in the following figure:

Weapons ⟷ Wars

Figure 4.8: Simple directed graph representing the mutual causal effect (reverse causation) of weapons and wars

It took me seven years to statistically solve this issue and isolate weapons' effect on wars. My statistical model revealed that, indeed, weapons imports increase the probability and intensity of conflicts, as well as the number of people fleeing the country (see Gallea, 2023, *Weapons and war: The effect of arms transfers on internal conflict*: `https://www.sciencedirect.com/science/article/pii/S0304387822001432`). I will talk more in detail about this paper in *Chapter 5*, when we discuss the solutions.

While I was still working on this research paper, at the end of my Ph.D, COVID-19 hit. Like many researchers globally, I could not focus on anything else. Everything was moving so fast. Countries and governments had to make decisions, and as you understand now, causal inference is key to making good, informed decisions. So, jointly with four other researchers, we started studying the effect of COVID-19 lockdowns on the spread of the virus.

Early in the pandemic, around March 2020, when almost the whole world was locking down, some people showed a positive correlation between lockdown stringency and the number of COVID-19 -related deaths. In this situation, both indeed affected each other. When the situation was out of control, the government would tend to implement stronger measures, and COVID-19 lockdowns almost certainly affected, in one way or another, the cases and related deaths. Again, given my skeptical mindset, I didn't approach this question with the willingness to prove that lockdowns were useful; I rather approached it with a neutral view and focused on the right statistical approach to properly answer the question.

Both examples illustrate particularly well the importance of causal inference. It is not just about saying "it works" or "it doesn't." Causal inference, if done properly, will allow us to precisely quantify the effect, which is necessary for decision-making. Most decisions, policies, and treatments have a cost. Hence, even if we have strong support for the sign of the effect, it is worth quantifying to ensure that the decision will have a net positive effect. Let me illustrate with the case of COVID-19 lockdowns. COVID-19 lockdowns had numerous costs: economic, social, and psychological. Hence, it was important to precisely know the benefit to assess whether it was worth it or not. Knowing that it reduced the mortality rate was not enough; we needed to know precisely how much. Our estimates revealed that COVID-19 lockdowns saved approximately 3.5 million lives after 100 days. It is important to note that this result is primarily driven by developed countries, as we didn't find any statistically significant effect for developing countries. The difficulty of respecting lockdowns in poorer areas might be one reason. This information is key to evaluating the decision to lock down countries or regions in the future.

As I mentioned earlier, the best way to detect reverse causation or simultaneity is to ask: "Could it be the reverse?". Note that both issues – "Is there something else?" to spot confounders, and "Could it be the reverse?" to question reverse causation – can coexist. In the two research examples I mentioned, I had to take both into account.

The last culprit questioning the causal claim is slightly different from the two we have just seen. Surprisingly, this issue can happen even if you initially measured or witnessed a clear causal effect.

External validity

Here comes the last tool of this section, which also takes the form of a question: "Can we extrapolate?" You might have very clean and clear causal evidence in one place, at one point in time, or with certain types of individuals, but we might not be able to apply those results in another context. Technically speaking, this is what we call **external validity**: to what extent are the results valid beyond the context studied? It is also important to define internal validity, which is the degree of confidence we have that a cause-and-effect relationship observed in a study is trustworthy and not due to other factors. Internal validity is always the starting point. Before even thinking about extrapolating a result, it is necessary to assess whether it is valid in the first place within the context observed.

External validity is critical because it answers the "So what?" question: can the results from this specific study be generalized to a broader population or context? Most of the time, we gather evidence by using the past to inform our future decisions. In business, we study the effect of our last marketing campaign to decide on the next one, or try to understand what caused a bad quarter to decide on future actions to reduce risks. In my current work on using causal inference to know what works and what doesn't work with AI implementation, this also helps to decide what to scale and which project to shut down. But this point is also valid for medical research, where we study rats in a lab or healthy patients in randomized controlled trials, but mainly care about people in real-life scenarios taking a drug, or in policy making, where we explore the impact of past policies to inform future decisions.

Now, let me illustrate the concept of external validity with potentially the best and arguably most terrible example I know. I have been teaching causal inference to experienced researchers in medicine and biology for quite some time now, and in particular, every year professionals preparing to lead animal research labs. When I started, the organizer told me that I should stress the importance of external validity, and in particular, with respect to sex differences. I was initially puzzled and took some time to understand. I was puzzled, not because I didn't understand the importance; I was puzzled because to me it was so obvious that I was shocked to realize that it was often not required or taken into account. The consequences of this issue are profound. Historically, medical research focused on male subjects. There are arguably two reasons for this disparity. The researchers wanted to protect the potential future mothers and their babies. Second, the menstrual cycle creates more variability and hence might complicate the analysis. But if research is conducted exclusively on men, it is important to ask whether we can truly extrapolate the effects to women. The fact that we know that there is a systematic difference that affects the outcome – the menstrual cycle – already foreshadows that the answer will often not be a simple: yes. Indeed, at the end of the 1990s, eight out of ten drugs withdrawn from the US market were removed due to side effects occurring mainly or exclusively in women (see Carey et al., 2017, *Drugs and Medical Devices: Adverse Events and the Impact on Women's Health*: `https://www.ncbi.nlm.nih.gov/pmc/articles/PMC5779632`).

This is a common issue when it comes to scientific research. Many people cite scientific paper results to make their point as an authoritative argument. However, scientific research is often done in a very specific context. Hence, it is worth questioning the context first and then evaluating the risk of extrapolation. If you want to extrapolate the results of medical treatments from males to females, you should consider whether the differences between the two groups could affect the effectiveness of the treatment. This is also valid if someone refers to data from the 80s to explain a current situation, or uses results observed in the US to predict outcomes in India. The results might be internally valid, but become questionable when applied in different places or times. It could even be on a smaller scale. You observe a marketing campaign working in one geographical region, but it might not be easy to extrapolate to another with a different culture or socio-economic factors. The efficacy of a campaign can also vary for different groups of individuals. As a reminder, as we saw in *Chapter 2*, this is one area where causal AI is particularly useful, as it helps to detect the heterogeneity of a causal effect, identifying in a data-driven way how subgroups might react differently.

Another type of extrapolation that might be problematic is related to the size of the treatment. Think about the effect of giving a dollar to a person living on under one dollar per day on the person's well-being. If you measure this effect properly, can we expect a proportional change if you give the same person 1 million dollars? Perhaps the initial dollars have a significant impact, but the marginal benefit decreases as more money is given. This is what we call diminishing returns.

We can also illustrate this concept with the example of the effect of the number of students per teacher on grades. You might find a study showing that a reduction to between twenty and twenty-five students increases the grade point average by 1 point. However, you might not be able to perfectly infer the effect of diminishing a class from ten to five students. Not everything is linear and can be linearly extrapolated. Maybe too few students affects their learning ability negatively at some point, if collaboration and interaction with peers help in some way.

The following illustration shows that within a certain range of values, the relationship between the two variables on the graph is almost identical (within the red square). However, outside this set of observed values, the effects are completely different.

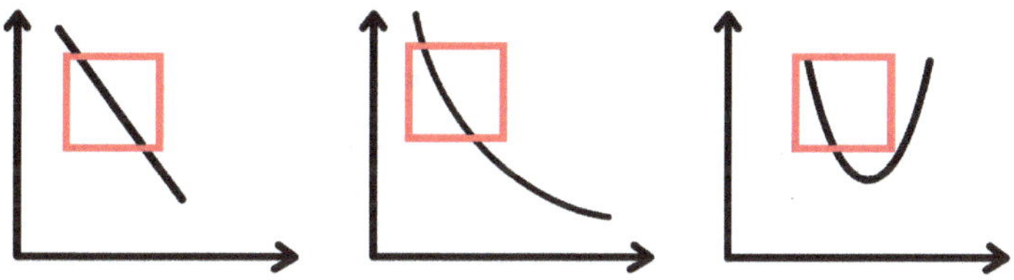

Figure 4.9: Three different relationships represented, which are similar within a set of values but not beyond

This concept is valid beyond scientific research. In industry, there is a growing use of causal inference methods to measure the impact of marketing or strategic decisions. A certain ad campaign done in winter might have a different impact during summer. Imagine an e-commerce platform selling designer furniture. Historically, its sales peak before Christmas, with numerous clients choosing to buy themselves a gift. This high-standing shop always promotes the most prestigious and expensive products during this period. Once again, the e-shop runs its usual strategy, and its sales fall. Indeed, the economy is slowing down, inflation is rising, and it is expected to keep rising without any clear sign that it will stop soon. Consumers might postpone expensive, compulsive buys of non-essential luxury goods. Maybe the shop would have benefited from rethinking the strategy, adapting to the economic landscape, and focusing on durable and accessible products.

Case studies

Now, let us apply our extended knowledge to the directed graphs we created at the end of *Chapter 1*. You can look at a directed graph and first identify which element is a confounder, a mediator, or just an external factor not affecting the causal relationship, and then reflect on whether reverse causation is a problem.

Case study 1: Cold showers and the immune system

What if we compare the health of people who take cold showers with those who do not take cold showers, as suggested at the end of *Chapter 2*? On the graph we drew at the end of *Chapter 1* for the first case study, identify the potential confounders, mediators, colliders, and external factors.

Suggested solution:

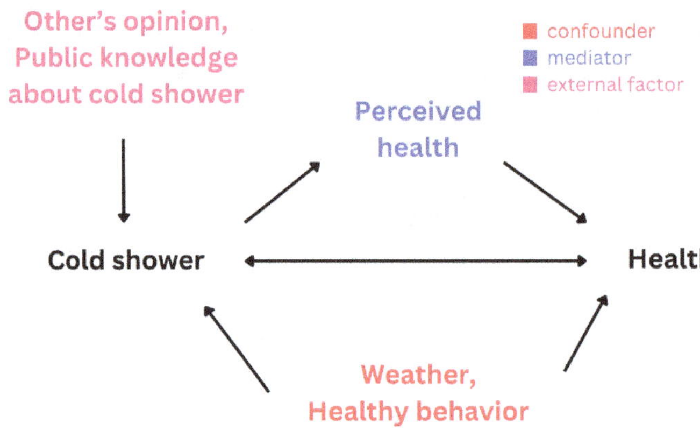

Figure 4.10: Directed graph representing the causal relationship between cold showers and health

The first question to ask is: "Is there something else affecting or explaining the correlation between cold showers and health?". The main problem here is that the weather and healthy behavior act as confounders. The two might affect the probability of taking cold showers and our health as well. Hence, if people taking cold showers are healthier on average compared to others, we don't know whether it is thanks to the cold showers or because they take other measures to be healthy (e.g., a healthy diet or regular physical activity); those effects are confounded.

The second question should be: "Could it be the reverse?". And yes, as suggested with the story of the old man swimming in cold water in this chapter's *Spotting reverse causation* section, being healthy might allow one to swim in cold water. But actually, some health issues might also be a driver to swim in cold water. There have been discussions about the potential benefits for people suffering from autoimmune diseases. While I suggested two potential opposite effects of health on the probability of taking a cold shower, it seems that, anyway, reverse causation is likely. Then, we can also notice that there might be a placebo effect represented by the perceived health node. This is a moderator as it simply represents an indirect effect.

Finally, others' opinions and public knowledge about cold showers might influence the probability of taking a cold shower, but not health directly. These are just other factors that do not affect the correlation between cold showers and health.

Case study 2: Marketing campaign for energy sobriety

Comparing the energy consumption in Switzerland in the months before an energy sobriety campaign started versus after could be misleading for several reasons, as suggested in the second case study discussion at the end of *Chapter 2*. Use the three questions that you have seen in this chapter to identify confounders, question reverse causation, and question external validity. For external validity, try to think about the validity of such a campaign (assuming it worked in Switzerland) in another country, and what could question the capacity to extrapolate.

Suggested solution:

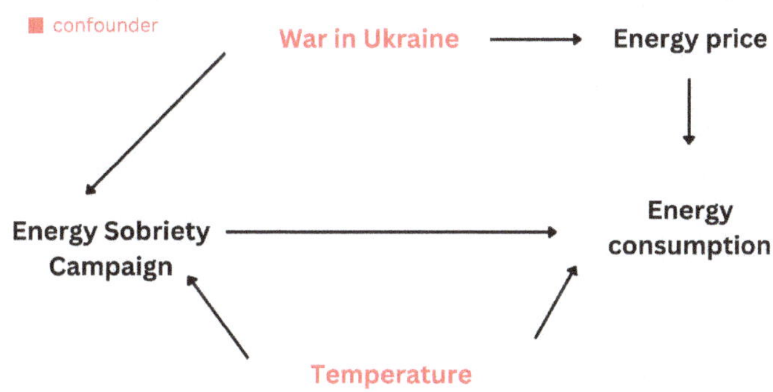

Figure 4.11: Directed graph representing the causal relationship of an energy sobriety campaign and energy consumption

This marketing campaign was triggered by the war in Ukraine, which affected energy supplies in Europe. The reduced supply created fear of shortages and price increases, hence potentially affecting energy consumption directly. People and companies anticipating a price increase might voluntarily lower their consumption to reduce their bill regardless of the campaign. Moreover, temperature affects the use of energy. More specifically, in Switzerland, heating systems consume a considerable share of the energy during the winter months. The months when the campaign was in place are also colder months, as it started in the fall. So, the effect observed on consumption when comparing simply before and after is confounded with the effect of the temperature and energy prices.

Now, would this energy sobriety campaign work as well in another country? It depends on the capacity to adapt. In Switzerland, a wealthy country, people were probably "over-heating" just for comfort and not survival. In such situations, you have more slack to adapt and turn down the thermostat with no severe consequences. In the same vein, after questioning the spatial extrapolation, we can question the temporal extrapolation. If the winter were particularly cold, it might be harder to save energy by reducing the temperature in buildings, reducing the effectiveness of such policies.

Case study 3: Air pollution and respiratory disease

As suggested at the end of *Chapter 2*, in our third case study, comparing people who live in highly polluted areas with people living where the air is clean might be misleading, mainly due to the difference in socioeconomic status. On the directed graph we drew at the end of case study 3, identify the confounders and discuss the risk of measurement errors.

Suggested solution:

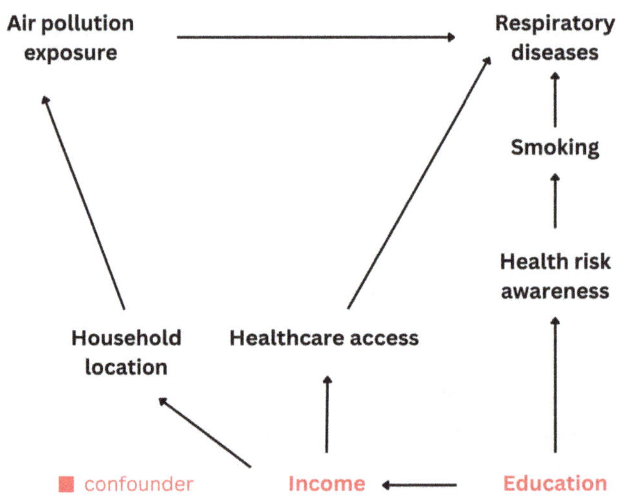

Figure 4.12: Directed graph representing the causal relationship between air pollution and respiratory diseases

More polluted areas might be populated by people from lower-income households on average. As people with more financial resources might have better access to healthcare services, which affects the risk of being sick or of suffering from severe complications, income is indeed a confounder (it is related to pollution exposure and to health outcomes as well). As you can see in the graph, I decomposed the pathway from socioeconomic status to respiratory diseases to be more explicit. However, the fact that there are multiple nodes along the pathway does not change the fact that it is a confounder, as it affects both the outcome and the cause of interest.

Respiratory disease prevalence might also suffer from measurement errors. We might primarily detect severe respiratory diseases with complications, where people require hospitalization, while undercounting milder cases. If people in lower-income areas are less likely to seek medical attention for less severe symptoms, then our measurement of the outcome (respiratory disease) would be systematically biased. This is a case of non-classical measurement error, as the accuracy of our measurement is correlated with the cause of interest (pollution exposure, which is linked to income), further complicating our ability to isolate the true causal effect.

Summary

Virtually everything is multifactorial, as we discussed in *Chapter 1*. Hence, yes, there is almost always something else to consider. In this chapter, we moved from understanding why causality is hard to learning how to spot the specific reasons a simple correlation can be misleading. You have now expanded your causal toolkit with three critical questions to challenge any causal claim: "Is there something else explaining this relationship?", "Could it be the reverse?", and "Can we extrapolate?". You learned to identify the main culprits behind flawed claims: confounders that offer alternative explanations, the problem of reverse causation, where the direction of the relationship is questioned, and the subtle issues of measurement error. Finally, we explored external validity, questioning whether a finding from one specific context can be reliably applied to another. This information transforms you from a passive observer into an active causal detective. Armed with this framework, you can now systematically "unpack the problem" behind a news headline, a business report, or a scientific study. You are no longer just suspicious of a claim; you now have a precise vocabulary and a structured method to identify the specific weaknesses in an argument, whether it's a hidden confounder like in the chocolate-and-Nobel-laureates example, or a risky extrapolation like applying medical findings from one group to another.

Having learned how to diagnose the common problems that plague causal claims, we are now ready to explore the solutions. In the following part, we will see how researchers and practitioners prevent these issues from confounding their results. In the next chapter, we will work on the Causal Mindset Framework.

Join our community on Discord

Join our community's Discord space for discussions with the authors and other readers: `https://discord.gg/G4YmB6DQvf`

Part 3

What Are the Solutions?

In the third part, we will climb the slope of enlightenment by discovering the methods to measure causal effects. First, we will learn how to qualitatively assess causal claims in real time using a structured framework, the **Causal Mindset**. Then, we will explore how to measure a causal effect when we can control treatment allocation through randomized experiments, and finally, how to infer causality when we cannot control treatment allocation and must rely on observational data (quasi-experimental methods). We will examine the strengths and weaknesses of all these methods. While virtually any method has limitations, we will see how one approach can address the problems of another, allowing you to choose wisely. By the end of this part, you will be able to evaluate the validity of a causal claim in real time using the *Causal Mindset* framework, understand the pros and cons of experimental and quasi-experimental methods, identify problems in causal claims using causal inference tools, and select the most appropriate method for your context.

This part of the book includes the following chapters:

5

The Causal Mindset Framework

Let's start the final push to escape the valley of despair. Now that we have understood the problems and how to identify them, we will cover the solutions in this third part of the book. This part will cover three types of solutions in three different chapters. First, we will see how to do a real-time qualitative assessment of causal questions. Second, we will discuss how to measure causal effects when we can randomize the treatment allocation. Finally, we will cover methodologies to measure causality when randomization is not possible.

In this chapter, we will see how to synthesize all the concepts covered so far and apply them with a practical methodology—the **Causal Mindset framework**—to real-world scenarios.

In most situations, we don't have the time or skills, or it is just plain impossible to collect data and do the analysis to answer a causal question. Arguably, due to this issue, causal inference is underused. Many people would completely disregard the methodology, seeing it only as a complex toolbox requiring long analysis. However, I witnessed that we could use causal inference concepts for a quick qualitative assessment of causal claims, and many technical experts in causal inference are doing this.

Realizing this fact, combined with my willingness to help society, pushed me to create the Causal Mindset framework, a structured and non-technical way to tackle causal questions. This is the core of my work and what I've been teaching to thousands of people over the years, allowing me to witness its transformative potential. I remember after a talk to a group of CEOs in the financial industry, one approached me afterward and told me, "*There is a before and an after. Your talk was eye-opening and made me see the world differently.*" After receiving this type of feedback numerous times, I decided to make this even more accessible and write this book.

In this chapter, we will cover the following topics:

- Learning about the Causal Mindset framework
- When to use the Causal Mindset framework
- Understanding the importance of domain knowledge in causal inference

Learning about the Causal Mindset framework

We will now introduce the Causal Mindset framework, a methodology that allows you to move from suspicion (*I feel there might be something wrong*) to informed analysis (*I know how to evaluate whether there is a serious problem or not*).

The framework is composed of these five steps:

1. **What is the control group or proxy counterfactual?** Identify the comparison being made and evaluate its quality.
2. **Is there something else?** Search for hidden factors (confounders, or reverse causation) that could be driving the result.
3. **What is the direction of the bias?** Determine whether these hidden factors are amplifying or attenuating the true effect.
4. **Refining and iterating**: Strengthen your analysis by seeking better comparisons or isolating subgroups if needed.
5. **Can we extrapolate?** Assess whether the findings can be reliably applied to a different context or to the future.

Before we go through each of these steps with real-life examples, let me remind you of the two roles of causal inference, as the framework will serve this dual purpose. First, causal inference allows us to assess causality from past events. You can collect data and use causal inference to assess what happened in the past. The first four steps of the method serve this role. Understanding past events is often just a means to an end. We typically need this information to make informed future decisions. Indeed, the second role of causal inference is to help you predict what would happen if you make one choice or another, allowing you to choose wisely. The last step primarily serves this second role of decision-making.

There are different levels of complexity and reliability. Applying the concepts of this method (the Causal Mindset framework) can be done in real time to get an initial sense, while an in-depth causal analysis can be conducted with data and statistics to precisely quantify effects. Causal analysis and the current framework are also about systems thinking.

It helps to see the multidimensional aspect of a problem and how one decision could backfire. This method will help to improve decision-making, but also assess whether further causal analysis with a technical tool would be useful or not.

Let me illustrate the method with a concrete example. Imagine a company questioning the effect of its new internal chatbot on productivity. The company noticed a 2% increase in productivity metrics (number of tasks completed) since the launch of their new tool. This 2% increase was measured by comparing the last three months with the policy, April–June, with the three months before, January–March.

This result is presented during a meeting, and quickly, two opposing views arise. One person claims that they should stop this as the benefits are very small and the employees might become lazy and lose skills. Someone else vigorously defends the policy, saying that even a 2% increase in productivity is rather useful and might result in a significant increase in income. This debate is multidimensional (income, environmental impact, psychological effects, long-term vs. short-term, etc.) and complex, but we can start by questioning the first piece of evidence that was brought to the meeting: can we really assume that the 2% increase in productivity was caused by the chatbot? Answering this question is necessary to make an informed decision and compare it to the other dimensions.

Step 1: What is the proxy counterfactual?

The first step is to think about the most basic concept of causal inference, the fundamental problem of causal inference. Remember, the ideal setup would be to have parallel worlds in order to have only one difference between the two worlds (the cause), with everything else being identical, allowing us to pin down the effect of interest. As it is not possible to have this perfect counterfactual (a parallel world without intervention or treatment), we have to find something that will act as the counterfactual. To do so, we aim to find a situation as close as possible to the parallel world. We are now applying in a practical, complete, and structured way all the tools we have studied throughout this book.

This first step, critically evaluating the proxy counterfactual, can often invalidate a causal claim at the outset. If questioning the comparison reveals a major flaw, the evidence can be dismissed without further analysis. You can often kill a weak argument at this very first step. In this case, you would need to look for a better proxy counterfactual or control group, as I suggested in the examples in the following box.

Additional example: e-scooters

Remember the example of a company offering e-scooter services in a city center? The company compared the emissions from driving a car to the emissions from using an e-scooter and concluded that e-scooters had a positive environmental impact since they are less polluting than cars. Their conclusion was based on the assumption that e-scooters replace cars. However, it seems that e-scooters more often replace walking and public transportation, which are more environmentally friendly modes of transport. Therefore, the choice of proxy counterfactual here is misleading and significantly impacts the conclusion. This first step, where we identify the proxy counterfactual, helps us identify this issue and propose a more appropriate comparison: for example, e-scooters versus public transportation and walking.

Additional example: Tesla car crashes

Let me illustrate this point with another case I witnessed once in the *Washington Post* about Tesla crashes. The article published in June 2023 was titled *17 fatalities, 736 crashes: The shocking toll of Tesla's Autopilot*. While the article claims that those numbers are shocking, you might want to question what the reference point is to support this claim. Sure, people died, and it is a serious matter. However, road crashes are common and are a leading cause of death in the world. So, to assess whether this is really *shocking*, it is important to find the right comparison, or proxy counterfactual. Later, the article uses data, trying to back up its case. Here is an approximate replication of the figure presented in the article:

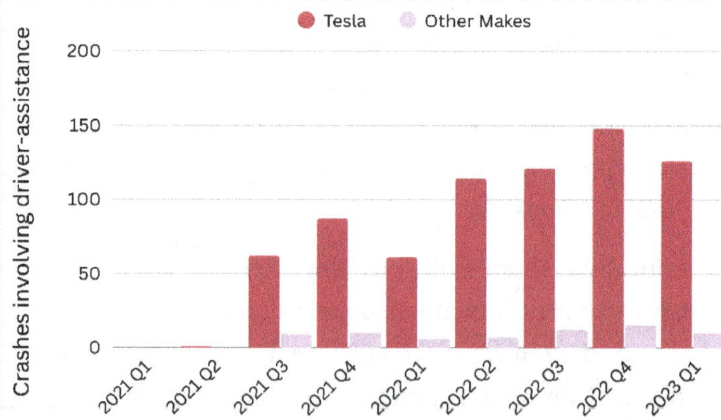

Figure 5.1: Number of crashes involving driver assistance per quarter in the US

> *Figure 5.1* shows the number of crashes by quarter in the US for Tesla compared to other makers from 2021 Q1 to 2023 Q1. The graph is accompanied by the following text: *Tesla's 'Full Self-Driving' and Autopilot systems have been involved in far more incidents than driver assistance systems from all other manufacturers combined.* The graph reveals two things. First, the number of crashes is increasing. Second, there are far more crashes with Tesla's cars than from other makers combined. However, both comparisons are uninformative about whether Tesla's Autopilot is becoming more or less lethal over time, and are therefore potentially misleading. The number of crashes increases during this period, but also, the number of Tesla cars on the road more than doubled during this period. Moreover, without information about the number of Tesla cars on the road compared to those of other makers, it is difficult to conclude anything from such a comparison. It would be more adequate to take into account the number of vehicles in circulation by looking at the crash rate to account for the market share and time variation.

The two additional examples revealed that this first step required finding more data to have a more reliable proxy counterfactual or control group. But what if the choice isn't so obviously flawed? Let's return to our chatbot and productivity case study. The company compares two different periods: the quarter before the policy (Jan–Mar) versus the quarter after (Apr–Jun). Is this a good proxy counterfactual, allowing us to make the right comparison? Perhaps not. An alternative would be to compare the same quarter from the previous year (Apr–Jun 2022 vs. Apr–Jun 2023) to control for seasonality. However, that comparison might be flawed due to long-term company growth or changing market conditions.

As highlighted before, domain knowledge is important and can help you choose between these two options. However, during this meeting, let's assume you can't change the data and still want to question the evidence presented in a robust way. Unlike the Tesla example, this comparison isn't immediately fatal; it's just imperfect. Our task, then, is to move to the next steps of the framework to determine how big the potential problem is and whether this suboptimal evidence still holds any value.

Step 2: Is there something else?

The second step is to question whether there are other factors related to our cause and outcome. This critical thinking is essential to ensure that we do not overlook other variables that could affect our analysis. The parallel world idea allows us to pin down the difference between what we observe and the counterfactual.

In addition, you can draw the causal directed graph in order to organize the relationships and identify which elements might be problematic to assess a causal effect. The main issue you want to detect is confounders: things that affect the cause and the outcome. Here is a representation of the most basic directed graph for this situation:

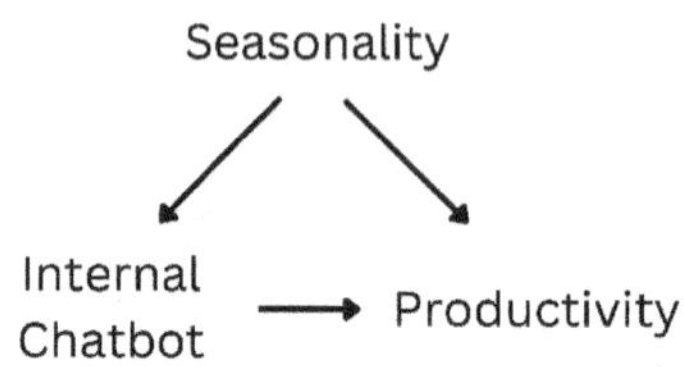

Figure 5.2: Directed graph representing the causal effect of the internal chatbot on productivity

In our example of the chatbot's internal use, is there something else (other than this GenAI tool) that is different between January–March and April–June? Maybe some specific events took place affecting one period differently than the other, or there is simply a seasonal difference with respect to the output produced in this company.

Let us assume that no particular events took place recently that are expected to affect productivity. However, the company has a well-documented seasonal change at the beginning of the year. January to March is usually the busiest period of the year, as many clients start new projects at the beginning of the fiscal year. Once we have identified potential confounders (here, seasonality), it is crucial to evaluate the extent of their impact on our analysis with the next step.

Another issue we've seen is reverse causation. Therefore, it is always valuable to ask whether the outcome could be affecting the cause. In this situation, a potential problem could be that the chatbot was introduced specifically to address pre-existing low productivity and dissatisfaction among workers. If so, any observed effect would be confounded with this underlying negative trend, which the chatbot may not be solving.

Using your domain knowledge, you can assess whether this is a credible threat and raise it in the meeting. For this example, let's assume you determine that this was not the case and that there is no significant risk of reverse causation.

This framework is about maximizing efficiency and practicality. Indeed, you might have other potential confounders and not just one, as illustrated here. But ultimately, you want to start with the biggest threat and already address this one if possible. Then, if you have other potential issues, and if time allows, as explained in *step 4*, you can iterate the process.

We have identified one major confounder: seasonality. How can we deal with this factor now? The next step allows us to assess quickly the severity of the problem.

Step 3: What is the direction of the bias?

Now that we have identified a major confounder, there is a last concept that you can use to efficiently assess the extent of the issue: What is the direction of the bias? Indeed, if the effect of the confounder goes in the same direction as what you observe, you have a big issue because it is hard to know whether the effect you observe is due to the confounder or to the cause you are interested in. When both go in the same direction, we call this an **amplification effect**. On the other hand, if the effect of the confounder goes in the opposite direction from what you are observing, you would have an **attenuation effect**.

The consequence of the two effects is vastly different. Often, you can live with an attenuation effect. An attenuation effect means that it attenuates the strength of the effect; in essence, it biases the estimation toward zero. This can happen with a positive or negative relationship. If anything, what you observe is a lower estimate of the true effect, as it is confounded with an opposite effect, reducing the observed impact. An amplification effect from a confounder is very problematic, as you would not know whether what you observe is only caused by the confounder, and hence whether your cause of interest plays any role or not.

This tool is particularly useful to quickly reflect and potentially move on. I have seen this approach used often during research seminars. Someone in the room would raise a potential issue questioning causality, typically a confounder that is not yet taken into account in the study presented. The author of the paper will reflect on the direction of the bias. If it seems that the confounder would attenuate the effect, the consequences are significantly less severe and do not entirely discard the results presented. On the other hand, if the discussion reveals that it is an amplification effect, it profoundly questions the results of the research.

Now, let's return to our example and apply the method. In our current situation, based on previous years, let us assume that there is usually a 5% decrease in productivity between the first and second quarters of the year. Hence, we are in the situation of an attenuation effect. The seasonal effect potentially attenuates the effect of the chatbot, as the observed change in productivity is actually positive for the first time in a few years during this period. So, in this situation, you could tell your colleagues that while both effects are potentially confounded, if anything, the real effect of the chatbot might be larger than the 2% observed.

Additional example: wage discrimination

The case of wage discrimination between men and women also illustrates the problem with amplification effects. If you listen to a debate on the topic, the main argument from people questioning the nature of the wage difference is that it is not discrimination, but that women, on average, earn less than men because they tend to work fewer hours, are often less willing to travel far or do risky jobs, and so on. All those are potential confounders as they affect the wage but also might be correlated with the gender of the workers. The main point why people refer to such confounders is that the confounders could explain the effect as they all have an amplification effect, allowing the detractors to claim that there is potentially no discrimination, as it might be fully driven by those other factors. Wage discrimination is a complex topic, and it goes beyond the scope of this book to present the findings in a detailed way. I recommend studying the work of the Nobel laureate in economics, Claudia Goldin, for some reliable analysis (for example, *Orchestrating Impartiality: The Impact of "Blind" Auditions on Female Musicians*: `https://www.aeaweb.org/articles?id=10.1257/aer.90.4.715`).

Additional example: marketing Campaign

Let me illustrate with another example I encountered. Recently, I was teaching data science for marketing at ISM University in Vilnius. I gave the students real marketing data from a company providing crypto-related products (designed to improve the security of wallets). One of the marketing campaigns was particularly prolific and was associated with a spike in sales. One of the student groups realized that during the same period, a major event took place: a leading platform of cryptocurrency trading went bankrupt (FTX). As the company was selling devices to secure crypto wallets, the spike in sales was most likely caused by this unique event, as suddenly, more people wanted to secure their crypto wallets. Missing this important factor could cause the company to overestimate the effect of its campaign and mismanage its investments. In this case of amplification bias, it is hard to say anything without further advanced technical analysis. The Causal Mindset framework, though, allows us to be very cautious with this correlation and highlights that it does not allow us to measure the impact of the concomitant ad campaign.

Once we assess the direction of the bias and the severity of the situation, it is time to assess whether we can move forward or need more data.

Step 4: Refining and iterating

At this stage, if you are in a situation where the confounder prevents you from concluding anything, you might want to start again with a different reference point (proxy counterfactual) or use more advanced solutions presented in the next sections. In the situation of the internal chatbot release, you could compare April–June with April–June from previous years, or you could compare the evolution of the productivity for workers who have access to the chatbot with those who didn't during the same period. Once you have another potential comparison, you can go through *steps 1* to *3* again.

In addition, to reinforce the evidence, you can go one step further and look for a situation where the confounder either has presumbaly no impact or has a stronger or weaker impact. In our situation, we mentioned that the source of the problem was the fact that there is an increased productivity during the first part of the year due to seasonality. What if some departments are not subject to this seasonal change? You could then focus on this subgroup to get closer to the true causal effect. This approach would likely increase the internal validity of your findings by providing more robust causal evidence, but it might simultaneously reduce its external validity, as the conclusions might not be generalizable to other workers in seasonal departments.

Once you have plausible evidence of a causal effect, you can move forward to the final step and see how reliable this evidence is for your decision.

Step 5: Can we extrapolate?

As we discussed at the beginning of this chapter, we often assess the causality of past events to guide future decisions. In medical research, knowing the effectiveness of a drug in a randomized experiment aims to know when and how to prescribe it in the future to patients. In marketing, uncovering the effect of a marketing campaign helps companies decide how to adjust their strategy for the next campaign.

Hence, once we have some causal evidence that we are satisfied with, we can try to infer about future consequences. However, the future is not the present, nor the past. Moreover, sometimes we might have causal evidence in a place, in a context, and we would like to use this for a different situation. As we have seen in this book, this should be done cautiously. In particular, we have seen in *Chapter 3, Why Is It So Hard To Prove Causality?*, that we should always ask whether we can extrapolate. Remember, by over-focusing medical research on males, numerous side effects have been discovered later on women while the drugs were on the market. This idea of external validity should always be questioned.

To question the external validity, you must reflect on the difference between the situation where you found causal evidence and the one where you want to make predictions. As discussed in the medical setting, if you want to use results on males to make predictions on females, you must study precisely the difference and the capacity to transpose those results. In particular, you can reflect on the differences that might affect the treatment effect.

One important consideration is that most of the time, we focus on measuring the average treatment effect. It means that we want to know what the effect of our cause on average is on some outcome. While the average is very useful and probably the most used metric to summarize information, it can be misleading, as we have seen in *Chapter 2, Causation versus Prediction*, when discussing heterogeneous treatment effects. The average can be misleading because it might hide treatment effect heterogeneity. This simply means that different subgroups within a population can react to the same treatment in very different ways. Imagine that you test a new drug for diabetes on a large sample of men and women. The results of the randomized controlled trial reveal that the effect is very strong and positive. However, it is possible that the positive average effect measured is driven by the results for a subpart of the population. What if it works only for younger people, while for older people it has no effect or even has a negative impact? Currently, this is a very hot topic in causal inference, in particular due to new methods allowing us to effectively measure this heterogeneity with causal machine learning (see *Causal machine learning for predicting treatment outcomes*: `https://www.nature.com/articles/s41591-024-02902-1`).

Back to our business example of GenAI use and productivity. Remember the context: we are in a meeting, and people are arguing whether we should continue with the chatbot or not. So here comes the question: Can we extrapolate our anecdotal evidence to the future?

In our situation, we observe a short-term effect, while if we keep the policy in place, we reflect on the long-term effect. People might improve their AI literacy and proficiency, hence getting more productive over time. On the other hand, some people might get lazy, and their productivity might go down. This is a known phenomenon with GenAI, often referred to as *falling asleep at the wheel*.

Here is another example: evaluating the effect of work-from-home policies on productivity. A recent study published in *Nature* using a randomized experiment to infer the causal effect found that hybrid working from home increased retention and job satisfaction without reducing performance (see *Hybrid working from home improves retention without damaging performance*: `https://www.nature.com/articles/s41586-024-07500-2`). However, note that your new tools allow us to question those findings as well: Can we extrapolate from this study to our company? This study was conducted in a Chinese tech company between 2021 and 2022. Although it is recent, there might be specific reasons why our company may not compare well with a Chinese tech company.

While we don't have all our answers, we have a precise framework to think about and question the assumptions. Moreover, it allows us to identify the issues and hence to potentially find solutions or at least know where to be careful. This Causal Mindset framework provides a powerful filter to evaluate the usefulness of a piece of information.

In our case study, we were able to move the debate forward by finding supporting evidence for the policy and decision-making. With the information at hand, there is no evidence that the policy should be stopped. However, we also realized that we are unsure about the effect in the future, which also allows us to make the decision to keep monitoring this multidimensional policy carefully. Finally, note that reverting this decision will also have potential consequences. It would not just be about getting back to the status quo, as people might be unhappy with this step backward. This itself presents another question you could now analyze with your newly acquired framework: What would be the causal impact of ending the policy?

Hence, to summarize this last step, you can question the external validity by identifying the differences between the period or place studied and the context where you want to extrapolate those results. For example, are there significant differences between the context of the past campaign and the future ones? For instance, if the market conditions are expected to change drastically (e.g., a new competitor enters the market, or a major economic recession), the effect of the marketing campaign might change. Second, you can consider heterogeneity by studying whether the effect is different for subgroups.

This was the final step of the Causal Mindset framework. In summary, this framework involves identifying issues using parallel worlds, questioning other factors, determining bias direction, refining through iterations, and making informed predictions. By applying this method, you will be more efficient, and it might become automatic. However, particularly in the beginning, it might seem complex. Hence, to help you in this process, you can check the case studies at the end of this chapter. In particular, reading *Case study 2* is very important to see a full real-life application.

The app

To enhance your understanding and application of the Causal Mindset, I've developed a companion app, accessible online. Access to the app is free and can be found at `https://www.thecausalmindset.com/app`. This app is designed to support your learning and make these tools even more accessible. It allows you to share a question, a statement, a document, or a graph, and it will automatically apply the method. This app is based on AI (large language model) and has been tuned to learn the Causal Mindset and apply it. The app will facilitate the application of the concepts and the learning process.

Furthermore, it seeks to empower you with these tools, fostering independence from the app itself as you learn by applying. Note that the app will not be able to know the true cause with certainty but it will help you reflect and evaluate risk.

But why is a tool for rapid, qualitative assessment so necessary? Why not always insist on a full, technically rigorous analysis? The answer lies in a crucial balance between practicality and rigor that every decision-maker must strike as explained in the next section.

When to use the Causal Mindset framework

This chapter might be surprising. The rest of the book shows how hard it is to pin down a causal effect and how rigorously we must work to do it correctly. While precision in causal inference requires careful work, most decisions often necessitate quicker approaches. As explained at the beginning of the book, being able to do proper causal inference might require a decade of studying and hard work, and then assessing precisely a causal effect might take a few years of scientific research. This timeline is not compatible with most of our decisions, when we need an answer today, this week, or maybe next month, for major decisions. This is arguably one of the main reasons why causal inference is underused.

Thus, the Causal Mindset framework offers a balance between rigor and practicality, enabling quicker, yet informed decision-making. The ambition of the Causal Mindset framework is to allow anyone to apply the concepts and improve the status quo, moving away from simple correlations and intuition, but also without taking years to answer one question. Despite my obsession with precision and unbiased estimation, the tools we have seen and apply in this chapter can make a profound difference while being very practical, not requiring a Ph.D in statistics or months of work. Sure, it will not be perfect, but it is far better than ignoring it completely.

Bottom line, the qualitative method presented here improves decision-making and critical thinking compared to working blindly with correlations or intuition by providing a method grounded in causal inference. However, it will not replace proper technical causal analysis but complement it. This qualitative approach can be used when the technical approach is not an option. If you are required to answer a complex question with important implications and you have the resources, you should invest in working with experts who would be able to apply the technical tools presented in the next two chapters.

This framework, whether applied quickly or in-depth, is not just a mechanical checklist. To use it effectively requires a critical ingredient that no statistical method can replace: a deep understanding of the subject matter itself.

Understanding the importance of domain knowledge in causal inference

In causal inference, domain knowledge (your deep understanding of the topic) is as critical as the technical knowledge of causal inference concepts. Indeed, to draw the graphs and think about bias and counterfactuals, you must know the topic well. To illustrate this, let's consider a recent policy decision.

Politicians in Switzerland want to reduce domestic meat production to reduce environmental impact. Indeed, meat is known to generate emissions and use land and water resources considerably more than legumes, another source of protein. However, if you draw your graph of cause and consequences, you would likely quickly realize that this policy could negatively impact the environment. The main issue is that reducing the domestic supply might increase imports, which is counterproductive for several reasons. First, the further the origin of the meat, the higher the emissions from transportation tend to be. Second, Switzerland has mostly small and circular farming operations that tend to be more environmentally friendly compared to massive intensive livestock production in other major exporting countries. Small farming operations in Switzerland allow farmers to use manure from cattle as fertilizer to grow crops. The absence of cattle might imply more production and the import of chemically produced fertilizer, which is polluting. Hence, there are several risks that this policy will negatively impact the environment unless it is accompanied by other measures, such as import regulations.

It is interesting to note that this particular case highlights a common problem with environmental issues. The metric used for the climate objective for agriculture in Switzerland is territorial emissions (emissions from the Swiss territory). Indeed, such policies might reduce territorial emissions. However, total emissions, and in particular Swiss consumption emissions (emissions from goods consumed in Switzerland), might go up. Such policies displace emissions from developed countries to developing countries. Using the wrong measures and ignoring causal analysis is likely to backfire. This is why education and spreading knowledge on causal inference are so important. This issue with the choice of metrics and a clear understanding of the problem using systems thinking will be at the core of *Chapter 8, Correct Answer, Wrong Question: The Importance of Choosing the Right Metrics*.

You can see that such reflection requires good domain knowledge to consider the interactions between the different elements. It requires knowledge of agriculture, pollution, and even economics. This is also why GenAI tools like my app can be valuable: they help users leverage broad domain knowledge to better reflect on causal risks. As a professional working on varied causal questions, I always start by deeply learning about the topic. If I work with a company, I always start by understanding their business, market, clients, and competition. I find such deep diving always fascinating and enriching.

Case studies

Now, let's apply the complete framework to the end-of-chapter case studies. This application will help to structure and go beyond all the steps we have done so far in one complete framework.

Case study 1: Cold showers and the immune system

Imagine that a colleague, such as Trent from *Chapter 1*, tells you during a coffee break, "*I've been taking cold showers for a month and I haven't been sick once. It definitely boosts your immune system!*"

This is a classic causal claim made in everyday life. Previously, we identified potential issues such as selection bias and confounding. Now, let's use the full Causal Mindset framework to structure our thinking and evaluate this claim in real time, as if we were in that conversation.

Step 1: What is the counterfactual?

The implied comparison is your colleague's health this month (with cold showers) versus his health in previous months (without them). As we have discussed previously, this is a very weak comparison. Trent might be subject to confirmation bias, and hence, he favors the evidence supporting a positive effect of cold showers, eventually ignoring the fact that he might be sick as he speaks with his colleague. In addition, this is a short period of time and concerns just one person, so there is also a high risk that any observed phenomenon could be explained by chance. As we have no further data to use, we can proceed to step two to highlight the weaknesses.

Step 2: Is there something else?

Trent seems to reject the fact that he is currently probably sick. Hence, we could also highlight the potential confounders we identified in our causal graph from the end of *Chapter 4*. Is it just cold showers, or is it the fact that your colleague is generally very health-conscious and has put in place other actions such as diet change, physical activity, or better sleeping habits? This "healthy behavior" is a classic confounder. We also identified reverse causation as a possibility: maybe being healthy is what allows him to tolerate cold showers in the first place. Imagine that he was weak and sick for months, but suddenly, he solved an underlying health issue and started having cold showers as he was doing better and wanted to keep improving his health.

Step 3: What is the direction of the bias?

Now that we have identified two major issues, let's question the direction of the bias. All the potential "healthy behaviors" might also improve health outcomes, and hence, they amplify the observed effect. This means the confounder (healthy lifestyle) or reverse causation both push the outcome in the same positive direction. This is a clear case of amplification bias. The perceived effect of the cold showers is likely exaggerated or even entirely explained by "something else."

Step 4: Refining and iterating

In a real conversation, you can't run a new study or collect more data. But you can use the first steps to gather more important information. You could ask a simple question to test the *healthy behavior* hypothesis: "*That's interesting! Have you changed anything else recently, such as your diet or workout routine?*". His answer would help you understand how much of the effect might be due to these other factors. This shifts the conversation from a simple claim to a more nuanced exploration of causes with a structured framework.

Step 5: Can we extrapolate?

Even if we assume that the cold shower had a genuine positive effect on your colleague, can we extrapolate this to you or other people? Perhaps it only benefits people who are already healthy, or maybe the effect is purely psychological (a placebo effect). In addition, a common discussion regarding cold plunges for women concerns their relationship with the menstrual cycle. There seems to be neither consensus nor solid scientific evidence, but it is certainly an important aspect to explore and consider when thinking about external validity in this case. The evidence, which, in this case, is extremely weak and comes from just one person's experience, has very low external validity.

By walking through these five steps, we move from a vague suspicion to a structured analysis, all within the space of a short conversation. We can conclude that while the colleague's experience is interesting, the causal evidence for his claim is very weak due to a flawed counterfactual and a high risk of amplification bias.

Case study 2: Marketing campaign for energy sobriety

Remember, in September 2022, Switzerland launched a national ad campaign to encourage energy sobriety in response to a potential energy shortage due to the war in Ukraine. Imagine you are a policymaker in a meeting, and you are presented with a simple before-and-after chart showing that energy consumption did not decrease after the campaign started. A colleague concludes, "*The campaign was a failure.*" Using your framework, you mention that the comparison is flawed as the use of *before* as a proxy counterfactual suffers from the confounding effect of seasonal change affecting energy consumption. So, when you arrive at *step 4*, you suggest to the data analyst team to provide new data taking into account the seasonality effect.

During the next meeting, the technical team proudly presents the following graph:

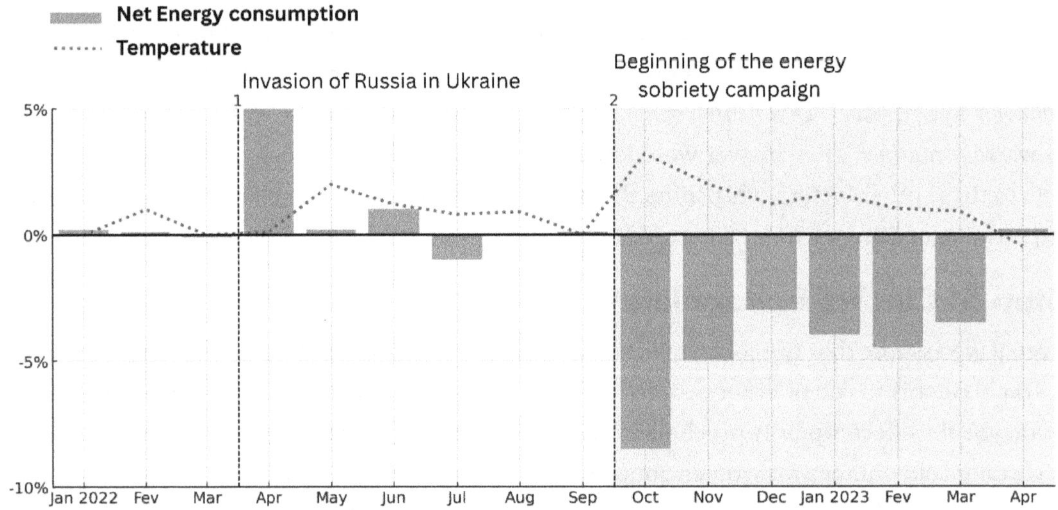

Figure 5.3: Net energy consumption and temperature in Switzerland. Both series represent the difference in percent from the average monthly value for the last five years

The graph represents a timeline with the net energy consumption in Switzerland and the temperature by month from January 2022 to April 2023. Note that this is an approximate representation of a real graph made by the Swiss Statistical Office and circulated online and in the news. The analyst made the following computation to take into account the issue of seasonality. They computed the average monthly value of temperature and energy consumption over the last five years. Then, they subtracted this value from the values for each month represented on the graph. Let me use an example to explain clearly how to interpret the numbers. In April 2022, the energy consumption in Switzerland was about 5% higher than the average energy consumption over the last five years' average energy consumption in April.

The graph reveals clearly a drop in consumption since October, once the campaign started. Everyone claps and concludes that the campaign was successful. Everyone except you. You look at the graph and start applying your five-step method. Note that this is exactly what I did online when seeing this graph. I started to discuss this piece of evidence with news reporters and policymakers.

Step 1: What is the counterfactual?

This comparison is incredibly more useful than before and reveals the monthly deviation from the average value for the same month, allowing it to capture average seasonality. The counterfactual is not trivial. It is a before-and-after comparison of the deviation from the average value for the last five years. While this is convincing and you probably really want to believe this, you still want to challenge the result as you learned to be careful and, in particular, suspect that confirmation bias could affect your willingness to accept these results.

Step 2: Is there something else?

Yes, there is *something else*; temperature is not perfectly captured. The graph also reports the deviation from the average monthly temperature with a dotted line, and you realize that this deviation is not zero. For example, in general, it was warmer than usual as the dotted line is above the 0% value.

Step 3: What is the direction of the bias?

A higher temperature in winter implies that there is a lower need for heating buildings, and hence, due to the weather, we could expect a reduction in energy consumption. This creates an amplification bias, and you know that this is a big issue, as now you are not sure anymore what drives the effect: the campaign, the weather, or both. Moreover, the more you look at this graph, the more you realize that the higher the temperature difference, the higher the energy saving. The values for April 2023 hammer the nail. The temperature was colder than usual, and the energy consumption was higher than usual.

Step 4: Refining and iterating

While everyone is still congratulating themselves, you stand up and start explaining the issue. The analyst did a great job capturing average seasonality. Moreover, the effect is sudden and strong, which represents a clear signal. However, it is still important to consider that temperature is a clear confounder, potentially amplifying the effect.

Such national marketing campaigns might cost millions. For such an important decision, representing large financial resources, it would be important to assess whether the benefits were worth the cost. To do so, you need more precise estimates and can ask your data team to use causal inference, and in particular, the methods presented in *Chapter 7, Quasi-Experimental Methods.*

Step 5: Can we extrapolate?

The main reason to assess this campaign is to inform future policies. However, we must consider whether the results will hold over time. With rising global temperatures, future winters may be milder, potentially reducing the energy required for heating. Therefore, the effectiveness of a campaign focused on heating could diminish over time. To investigate this more precisely, you could study the heterogeneity of the campaign's effect. For instance, comparing the impact in colder regions versus warmer regions within Switzerland would provide clues as to how much of the campaign's success was tied to heating behavior, and how its effectiveness might change in a warmer future.

Case study 3: Air pollution and respiratory disease

A newspaper headline claims: *Study Shows People in Urban Area A Have 30% Higher Rates of Respiratory Disease than in Rural Area B, Proving the Impact of Air Pollution.* This causal claim triggers your Causal Mindset, and you start applying the framework.

Step 1: What is the counterfactual?

The counterfactual being used is the health of people in rural area B, which is meant to represent what the health of people in urban area A would be if the city had clean air. This is a problematic comparison from the start. We cannot assume that the only relevant difference between these two populations is the quality of the air they breathe. In this case, the time period is identical, but the spatial dimension is not.

Step 2: Is there something else?

Income is a major potential confounder we identified in previous chapters. Highly polluted areas are often correlated with lower average incomes, while rural areas may have different demographic and economic profiles. As we mapped out in our causal graph, income can affect both the "cause" (where people live, and thus their pollution exposure) and the "outcome" (health, through factors such as smoking habits and access to healthcare). We also have to consider non-classical measurement error: if lower-income individuals in the polluted city are less likely to report or seek treatment for mild respiratory issues than the rural population, our data will be systematically biased.

Step 3: What is the direction of the bias?

Lower income is generally associated with poorer health outcomes. If polluted areas also have a lower average income, then this confounder creates an amplification bias. The observed 30% difference in respiratory disease is likely a combination of the true effect of pollution and the pre-existing health disparities linked to income. However, the non-classical measurement error we mentioned would likely create an opposite attenuation bias, as under-reporting of health issues in the more polluted, lower-income areas would make the observed effect appear smaller than it truly is.

Step 4: Refining and iterating

In this situation, the estimation is noisy. We have two potential biases pushing in opposite directions. Do they cancel each other out? Is one stronger than the other? We don't know, and therefore, it is difficult to conclude anything from this data alone. To move forward, we would need to analyze the income distributions of both areas to better assess the magnitude of the confounding effect.

Step 5: Can we extrapolate?

This five-step analysis shows that the initial headline, while dramatic, is based on very weak causal evidence. Given the severe internal validity issues, it is not worthwhile to even discuss the potential for extrapolating this result.

Summary

In this chapter, you learned about the five core steps of the Causal Mindset framework, a practical and qualitative method to challenge causal claims in real time. We walked through how to identify the proxy counterfactual, ask "Is there something else?" to spot confounders, assess the direction of the bias, know when to refine and iterate, and question whether results can be extrapolated. You saw how to apply this framework to dissect real-world examples, from corporate policy decisions to misleading news headlines.

The Causal Mindset framework helps to reflect quickly and evaluate risk in various contexts, from everyday life discussions to important strategic decisions, or when listening to the news. It helps to move the discussion forward and make further decisions. In situations where you don't have time or resources to work with the data and an expert in causal inference, this will improve the quality of the insights. That being said, it is imperfect, and if the decision is important, it would be recommended to seek expert analysis on the question to properly capture the causal effect. The goal of this framework is not to replace expert technical work, but it is made to fill a gap: the countless situations where we can't do a full technical causal analysis.

You now have the ability to rewire your brain with these tools and improve your thinking in real time. Over the years, I witnessed many people who made this change. After being exposed to this method in class, during conversation, or on social media, they came back with examples where it was useful in their personal life and dramatically improved their critical thinking. Now, it is your turn.

Now that you are equipped with a framework for the rapid, qualitative assessment of causal claims, the next two chapters will explore the more technical solutions. The next chapter will present the "gold standard" for measuring causality: randomized experiments.

6

Randomized Experiment

We are now slowly but surely climbing the slope of enlightenment and escaping the valley of despair. You have acquired *The Causal Mindset Framework* and are now able to assess the validity of a causal claim in real time. This framework also allows you to identify whether you need to go further.

When a decision has important consequences in business, politics, or research, you need to quantify precisely the size of the effect. In this chapter and the next, we will dive into the methods that allow us to do just that. We will begin with the so-called *gold standard* of causal measurement: randomized experiments.

In this chapter, you will learn why randomization is crucial to pin down causality. This is the same methodology used to approve life-saving medicines and is the engine behind the A/B testing that drives growth at major tech companies. Understanding how to design and critically assess these experiments is essential for any leader who wants to move beyond guesswork and make high-stakes decisions based on the strongest possible evidence.

In this chapter, we will cover the following topics:

- Designing a credible experiment
- Assessing randomized experiments' weaknesses and limitations

Designing a credible experiment

The first chapters highlighted that, ideally, to measure a causal effect, we would need to observe parallel worlds, and this impossibility created numerous issues: confounders, reverse causation, and non-classical measurement errors. In short, the main problem was that many things happened at the same time, and hence it is impossible to know what caused the result. The most popular way to address this issue is to set up a **randomized experiment**, often referred to as **randomized control trials** in scientific research or as **A/B testing** in the business world.

Randomized experiments are not solely taking place in academia or in the pharmaceutical industry; they are now widespread in other industries where they are referred to as A/B testing. Gregory Crawford, Zalando's chief economist, said in an interview that some tech companies run more than 30,000 A/B tests per day (`https://www.youtube.com/watch?v=sA7GtNImP2U`). In the same interview, Gregory explained that there's an "*aha moment*" when leaders realize that, instead of relying on beliefs, running A/B tests lets them see the effect in real data, changing how they approach decisions. In another video, Ian Yohai, Netflix's director of growth and payments, said "*At Netflix, we have a strong culture of experimentation, which means that all our product changes are tested*" (`https://www.youtube.com/watch?v=WRGW6xHLy3k`). Actually, you have probably been subject to numerous randomized experiments already without knowing it. Typically, if a certain social media platform would like to test a new function, or change the position or color of a button, they can randomly change the page for some people and keep showing the usual page to others. Then, by studying the behavior of the users, they would decide whether the change is beneficial or not, and hence whether they want to roll out the new version or keep the old one. This strategy is commonly used in marketing as well. Companies can create different variants of an advertisement and see which one is the most efficient.

The idea of a randomized experiment is simple: you select a group of entities (typically individuals) and randomly assign each one to either a treatment group, which receives the intervention (e.g., a drug), or a control group, which does not (or receives a placebo). If everything is done correctly, the only systematic difference is the treatment, and hence it allows them to pin down precisely the causal effect of the treatment. This situation is very close to the parallel world situation and creates a powerful proxy counterfactual. People do not decide whether to take the treatment or not; often, they don't even know whether they took the treatment (see the concept of "blinding" and placebo later in this chapter), and hence it drastically reduces, if not completely, the risk of having a confounder or reverse causation.

This is explained in the following figure:

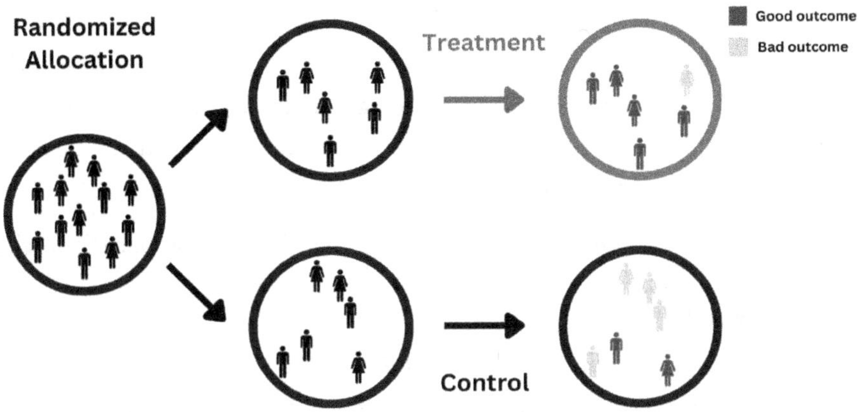

Figure 6.1: Conceptual representation of a randomized experiment

Randomization allows us to break the link between potential confounders and the treatment. Imagine that you study the effect of seating position on academic performance. You observed in the past that students seated in front have better exam results. But, using the Causal Mindset, you ask, "*Is there something else?*". And, the answer is yes; more motivated and hard-working students tend to sit in the front rows. Motivation would be a confounder as it influences both where a student sits (front vs. back) and the outcome, namely, academic performance. Indeed, if a student is more motivated, they are likely to pay attention, do homework, and perform better, regardless of where they sit. Now, imagine that you decide to randomize seats at the start of the term (e.g., by lottery or rotating chart). Some students will be assigned to the front and others to the back. If this happens, motivation is no longer a confounder for seat position: motivation still affects performance, but it is no longer related to seat assignment. The difference in performance between front- and back-assigned students can then be interpreted as the causal effect of sitting in front. This situation is depicted in the following figure:

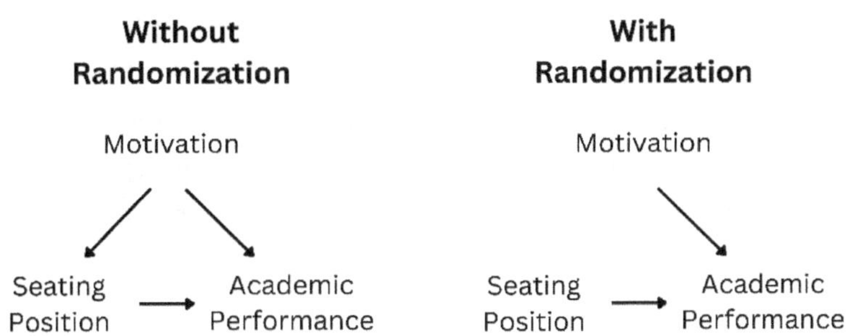

Figure 6.2: Directed causal graph representing the relationship between seating position and academic performance with and without seating position randomization

Of course, it is not that simple. Hence, we will now dive into this methodology, its steps, using examples from research and industry to have a profound understanding.

In what follows, I will present several medical research examples, as most people will have such examples in mind as the gold standard. But as we are going to see, numerous things can go sideways even in such controlled environments, and hence these lessons are equally valuable for other setups, including in the business world.

Let's start from the beginning: how we select a sample, and the implications of this choice.

Selecting a sample

The first step of a randomized experiment is to obtain a sample, assuming that we have already defined what we want to test. A **sample** in this context refers to a set of observed entities (e.g., lab rats, individuals, cells, users of an app, etc.). This sample should be representative of the population of interest. For example, if you want to test the effect of a drug for the whole population, you would not select a sample composed exclusively of males, as it might prevent knowing the efficacy for females (exactly as it has been done too often in practice, unfortunately). The first important consideration aims to avoid having a sampling bias.

Sampling bias arises when the individuals selected for a study are not representative of the population of interest, and hence, the conclusions cannot be extrapolated to this population. A famous cartoon illustrates this idea perfectly. Imagine that you submit a survey to your employees asking them to choose either *Yes, I love responding to surveys* or *No, I toss them in the bin*, and you get the result that 99.8% love responding to surveys. If the survey was voluntary, this impressive percentage is likely due to sampling bias. The result doesn't reflect the entire group's opinion; it only reflects the opinion of the small subset of people who answered the survey, with an over-representation of people who do enjoy answering surveys. Hence, it is not representative of the whole group. This specific type of sampling bias is called **volunteer bias** or **self-selection bias**.

Often, volunteers are not representative of the whole population, so it is always necessary to assess this risk. A low response rate in a survey, for example, signals a high risk of this bias. To address the issue, it is important to study the characteristics of the sample obtained to see how it differs from the population of interest. This is why collecting demographic data is so useful, as it allows us to check whether the observed characteristics of our sample match those of the broader population.

I stress "observed" characteristics because unobserved traits can also play a role, and these are much harder to assess. Imagine that you study the impact of cold showers on health by recruiting only volunteers. You might find that their demographics match the general population, but what about unmeasured psychological factors? For instance, what if these individuals are simply more mentally resilient? This unobserved trait could affect both their willingness to participate and their health outcomes, making it difficult to extrapolate your findings. Your domain knowledge can help to evaluate this risk.

There are practical steps to reduce this problem, particularly with survey data. To increase the participation rate, you can send reminders, make the survey shorter, or even make it mandatory. The response rate itself provides a first assessment: a 90% response rate is generally speaking a positive signal, but a 10% rate significantly increases the risk that your sample is unrepresentative. Even with a high response rate, you must check the demographics. What if 90% of your clients answered, but you missed the 10% of key clients who represent 80% of your sales? Being aware of and actively trying to mitigate the risk of sampling bias is a necessary step in designing a credible study.

The issue of sampling bias questions the external validity of the results, the capacity to extend the results to a larger unobserved population. As we have seen in *Chapter 4*, and in the last step of the Causal Mindset framework, we should ask "*Can we extrapolate?*" to question external validity. As long as the rest of the study is done correctly, you could find a causal effect even if your sample is not representative of the bigger group of interest. Sampling bias does not question the internal validity (the validity of the results within the setup of the observed data). This is central to remember and something I often even stressed when discussing with experienced researchers.

This concept of sampling bias is so important and often disregarded for cost reduction (focusing on an easily accessible sample) that a famous paper has been published in *Nature* on this subject, titled *Most people are not WEIRD* (`https://www.nature.com/articles/466029a`). The issue highlighted in this paper is that research relies extensively on **WEIRD** subjects, meaning people from **Western, Educated, Industrialized, Rich, and Democratic** regions. The paper makes the point very clear: those people are a minority on the face of the earth and not representative of the world population for various reasons.

Finally, in numerous situations, we can benefit from using a stratified sampling method. This method ensures that we select a sample with a similar distribution of the characteristics as the population of interest. Imagine that you want to launch a new product in Switzerland. Despite being a small country of 8 million inhabitants, Switzerland has 3 different regions speaking different languages: Swiss-German, French, and Italian.

Those differences in languages come with numerous other cultural differences. Hence, it is worth considering having a stratified sample, which is a sample with a representation of each stratum of the population proportional to its share in the Swiss population. This methodology allows us to ensure that the average difference between the treated and control group after randomization reflects the potential average effect for the entire Swiss population.

A different problem—selection bias, already mentioned earlier in this book—questions the causality of a result. Selection bias means that the treatment is not randomly allocated. It means that there are other systematic differences between the two groups. Imagine that you do not randomly force people to take cold showers, but just compare the health of people taking cold showers with those who do not. People who take cold showers voluntarily might put a lot of effort into being healthy, and hence, they might be healthier on average for various reasons aside from cold showers. This is why the treatment allocation is central.

Treatment allocation

Having established a representative sample, the next step of a randomized experiment is to randomly allocate the individuals to the different treatment groups (e.g., placebo or treatment). The random allocation aims to prevent the problem of confounders. By randomly allocating the individuals between the different treatment groups, you prevent having systematic differences between the treated and control groups, except for the treatment itself, allowing you to pin down its causal effect. Indeed, the individuals in both groups will not be identical, but if the randomization is made correctly, they should have a similar distribution of their characteristics (e.g., age, sex, weight, height, origin, etc.). Note that there are always differences because we don't have parallel worlds, but those differences are not problematic as long as they are not systematic, as it would not be confounded with the treatment effect, as we have seen before with the seating position in class and academic performance.

To deepen our understanding, let me use another case study. Imagine that you want to test the effect of giving access to ChatGPT on the output of your workforce. You select 50 people, and you give them all access to ChatGPT at the same time, in December, and you realize that their output was reduced during this period. Well, here, using the Causal Mindset framework, you could suggest that December is usually a month with lower productivity, and hence the effect of ChatGPT would be confounded with this seasonal effect.

Assessing this limitation, another approach could be to compare the output between workers who use ChatGPT voluntarily and those who don't in December to have the time dimension fixed. However, this approach could also be questioned within the framework. Those who use ChatGPT might behave on average differently; maybe they are more motivated, they are people constantly looking for solutions to be more efficient, they are more comfortable with technology, and hence, the effect of ChatGPT could be confounded with the effect of their skills and motivation.

This is why randomization is handy. To avoid those problems, you randomly allocate the 50 workers to 2 groups. You give access to ChatGPT to one group (treatment group) and you don't to the other (control group). Now, both groups of workers are working at the same time, which prevents the problem of seasonal changes. Second, the individuals are allocated randomly, and hence, the probability of having motivated employees in both groups is the same. However, note that here, you randomize the *access* to ChatGPT but not the *use*. People could have access and potentially refuse to use the tool. This does not affect the causal claim; it only affects the conclusions and interpretation of the results.

The impact of ChatGPT on productivity: evidence from randomized experiments

Understanding the impact of generative AI on work is central to navigating the current AI revolution. There are countless heroic claims online that generative AI boosts productivity and will destroy jobs, yet many projects in the industry do not deliver on their promises.

To answer this question, researchers, in particular from Harvard Business School, are doing randomized experiments. Specifically, they conduct "lab-in-the-field" experiments. This means they go into companies and observe workers in their natural professional environment, tackling real and concrete tasks. This approach increases the study's external validity.

The first, and arguably one of the most famous papers on the impact of generative AI on productivity, at the time of writing this book, is titled *Navigating the Jagged Technological Frontier* (`https://papers.ssrn.com/sol3/papers.cfm?abstract_id=4573321`).

This randomized experiment, conducted by top scholars with Boston Consulting Group, revealed that consultants working with ChatGPT worked faster, completed more tasks, and achieved, on average, a 40% increase in quality compared to those without ChatGPT.

But most people ignore or downplay the other finding that is the paper's main point: the jagged frontier. Sometimes it works; sometimes it does not. On one carefully designed task, consultants using ChatGPT performed 19 percentage points worse than those without it. Interestingly, the quality of the recommendations (presentation and structure) was higher, even when the answers were wrong, for those who used ChatGPT.

The second experiment was done in collaboration with Procter & Gamble and focused on a product innovation task and teamwork. While they also found an increase in performance for individuals or teams with generative AI compared to those without ChatGPT, one of the most original results was that using generative AI improved the self-reported emotional response (see *The Cybernetic Teammate: A Field Experiment on Generative AI Reshaping Teamwork and Expertise*, `https://papers.ssrn.com/sol3/papers.cfm?abstract_id=5188231`).

While the definition of the treatment and control group seems obvious, there are actually often many potential options. In *Chapter 3*, we discussed the concept of **counterfactual**. The choice of a proxy counterfactual or control group affects the conclusion of a causal question. This is actually a key question we asked in research when discussing causality: What is the counterfactual? Sometimes, you want to compare your treatment to a placebo, which is what we call a negative control, or eventually to a positive control, an alternative treatment. When I teach causal inference and biostatistics to researchers doing research with animals, this is a common discussion. In this context, you can compare a treated group of animals receiving a new treatment with a control group receiving a placebo. This would allow us to measure the incremental benefit of receiving treatment versus no treatment (placebo). However, if there is already an efficient treatment on the market, a potentially more important question is: Does the new treatment do better than the current standard of care? In this situation, you might want to compare the new treatment to the old one, which changes the counterfactual and the conclusion of the research. This is valid in every domain. Imagine that you want to evaluate the effect of a certain sports program, comparing doing nothing or doing another program, which leads to different observations. In marketing as well, do you want to compare the conversion between two new variants of an ad or the difference between the new versions with the previous old one? Sometimes we might not take any action, and the status quo used in the control group with a placebo or no action makes sense. However, often, we consider different options, and inaction is not one of the choices. Note that in marketing, A/B testing works well online, where we can track individual reactions. However, for a larger-scale effect, for example, with billboard campaigns, we need other methods to measure the impact. These methods will be presented in the next chapter.

As I said earlier, this book is not about superficial knowledge. Many researchers that I've met would stop at the first stage of randomization between treatment and control without realizing that other elements of the research were systematically different between the two groups as well.

Randomize everything that can be randomized

In *Chapter 3*, we saw several situations where the failure to randomize aspects other than the core treatment created problems.

First, recall the example of different experimenters administering the treatment. An inexperienced intern might mishandle an animal, causing stress and potentially slight physical harm, while a highly experienced researcher provides the treatment with exceptional caution. The resulting difference in care might create systematic differences in the general well-being of the two groups, which could become confounded with the effect of the treatment itself. This situation could be avoided if both individuals provided the treatments randomly (and, of course, if the intern is supervised).

The second example involved lab rats housed in different locations. If the placebo group is placed right under the AC where it's colder, or closer to a door with more activity, these environmental factors could affect the animals in ways that become confounded with the treatment effect. For example, the animals might be less active during the day (more tired) due to these systematic differences, which would be problematic if you are studying insomnia treatment and diurnal activity is an outcome of interest.

Let me take a final example. A research team compares three different surgical interventions on rats. To facilitate the process and reduce costs, the team performs surgical intervention *A* in the morning, *B* in the afternoon, and *C* in the evening. While this might be convenient, the surgeon might grow tired as the day goes by, negatively affecting the quality of the later interventions. Because the time of day is not randomized, this creates systematic differences confounded with the effect of the surgery type.

Of course, you can extrapolate those examples beyond a lab. If you want to measure the effect of different structures of a training course, but the speaker or the location is different, what is driving the outcome: the training structure or those other factors?

These examples reveal that systematic differences can arise from experimental conditions, time, or space, even when the treatment itself is randomized. It shows that everything that can be randomized should be randomized to avoid systematic differences between the groups. This is not trivial, and often financial or technical constraints prevent it from happening.

In the case of surgical interventions, for example, you might have two different surgeons who are experts in different methods. If the surgeon performing the intervention is fixed by their expertise, the surgeon's skill (a potential confounder) might be confounded with the type of intervention. Your domain knowledge is central to assessing and addressing such issues. With your expertise, would you say this is a high risk or not? Do you have any evidence that the two surgeons deliver work of different quality? Can you observe the quality of the surgery to ensure that the standard of care was respected in both groups? When everything cannot be randomized, domain knowledge and additional metrics help to mitigate the problems discussed in this section.

Even in a perfectly randomized environment, there are other potential issues that could arise and question the results. A participant's beliefs and expectations about a treatment can dramatically influence the outcome, a phenomenon we address through the use of placebos and blinding.

Placebo, nocebo, and blinding

The placebo effect is a well-documented phenomenon where a patient's symptoms can improve simply because they believe they are receiving an effective treatment, even if the treatment is inert. The nocebo effect is the opposite: negative expectations affect the outcome negatively. Hence, most of the time, it is essential to know whether the effect observed is just a placebo effect or a real treatment effect. To do so, a control group usually receives a placebo, which is something that looks like the treatment but has no actual effect. In medical research, this is fairly simple most of the time; you can have a pill, a cream, or an injection that contains no active substances.

However, beyond the world of medical research, it is becoming more complex. I have been working with companies to run randomized controlled trials to measure the effect of a specific type of training. In this situation, it is impossible to find a placebo training. Even if you manage to provide training that has no effect, people might perceive that it is useless, hence affecting them, their reaction, and even their stress level if they spend hours doing something that has no value on top of their work. In a situation when you don't have a placebo, it is central to assess the potential consequences, in other words, the placebo effect.

Using a placebo treatment prevents participants from knowing which treatment they receive. This equalizes expectations across groups, so placebo and nocebo effects affect both groups similarly and do not bias the estimated causal effect. This is what we call **blinding**. Ideally, research should be double-blind, meaning that neither the researchers nor the subjects know who received what during the study. Indeed, researchers themselves can be affected in their judgment, the way they behave, or how they evaluate a treatment's effect if they know who was allocated to which group.

Imagine you advocate for using ChatGPT to boost your team's productivity. Your superior asks you to prove its efficacy by evaluating the output from employees who were allowed to use AI, compared to those who were not. If you know who used the tool, your judgment in assessing the quality of the output could be biased, subconsciously favoring the work produced by the ChatGPT group.

However, in many situations in the business world, we have a very high level of blinding. Major companies constantly run A/B test changes on their website and apps without the users knowing. You are probably subject to such experiments numerous times every day without knowing it.

However, in some situations, it is impossible to blind people. In medical research, it is virtually impossible to blind surgeons during an intervention (e.g., if you compare two different surgical interventions) or in the industry with the training I mentioned before.

When it is impossible to blind, when blinding is partial or compromised, it is important to focus on reliable metrics: measures that can be observed precisely, that cannot be manipulated, or that are not subjective. In the example of ChatGPT, an assessment of the quality might be subjective, while the quantity of output, if clear rules are set, would be more objective. Read *Case study 1* at the end of this chapter for an applied in-depth example discussing this issue when blinding is not possible.

Blinding is not the only challenge, and considering the effect that the treatment might have on other groups is also central.

Preventing contamination and spillovers

An additional important condition to have a clean randomized experiment is the non-contamination principle, also known as **spillover**. There is a technical term to describe this in causal inference: **Stable Unit Treatment Value Assumption** (**SUTVA**). In the parallel world metaphor, the two worlds are not in contact. This is what we call **non-contamination**. In other words, each group should not be able to "contaminate" the other. In my research project on the effect of training on the psychological health of workers, it was important to reduce the contact between the groups. Imagine that after the training, the workers all went for lunch together and the treated group shared all the tips and what they learned with the control group. While this networking effect or spillover effect could be the subject of research, it should not affect the control group, else it would prevent properly measuring the effect of the training.

I have also heard another example from research. A team of researchers tested the effect of financial support in a poor area. But one of the observed groups would not receive any financial support. This could create clear issues of contamination. In this case, the control group (no financial support) is potentially affected negatively by the fact that other people in their neighborhood would receive support. This might discourage them and create frustration, which could affect the outcome of interest (e.g., school achievements). This type of problem relates to ethical considerations that should be assessed when considering running such experiments.

In the business world, a major challenge arises from the fact that users often visit a website or app multiple times during an experiment. This requires us to carefully track users for several key reasons. First, a common assumption in experimental design is that each observation is independent. If the same user is counted as a "new" visitor each time they arrive, this can violate that assumption and distort the results. Researchers must either ensure each user is observed just once or explicitly account for repeated observations in their analysis. Second, it is crucial to ensure stable assignment: each user must consistently see the same version (or "variant") of the experiment. If you go on an app and the position of a button is constantly changing, or a new feature appears and disappears, the inconsistent experience will likely frustrate you and affect your behavior in ways that contaminate the test. This stability is typically achieved by using cookies to track anonymous visitors or by linking the assignment to a user ID for logged-in customers.

As we have seen, the treatment might affect the control group. An almost opposite problem is the fact that sometimes the individual in the treatment group does not take the treatment, which is called **non-compliance**.

Non-compliance

What happens when you run a perfect randomized experiment, but people don't follow the instructions?

In some settings, the experimenter allocates individuals to different groups but can't ensure that they actually take the treatment. Imagine we are testing a new sleeping pill. We randomly assign 1,000 people to the treatment group (who are told to take the pill every day) and 1,000 people to the control group (who get a placebo). However, many people forget to take the pill. Hence, what you observe is not the effect of the pill; it is the effect of the recommendation and availability of the pill. In the absence of further information, you can still compare the two groups. However, instead of observing the effect of the treatment, you observe the **intention to treat** (ITT).

This might seem strange. Why include people who didn't take the pill? Because ITT answers the real-world policy question: What is the effect of *offering* this treatment?. For a policymaker or a business leader, what matters is the effect of launching a program in the real, messy world where compliance is never perfect. ITT preserves the benefit of randomization and gives an honest, real-world estimate of a policy's effectiveness.

This is also present in the business world when you give out vouchers. Not everyone will use the voucher they receive. Similarly, when studying the effect of providing a chatbot to a group. You might be able to track who uses it. However, the impact of the decision is rather reflected by the ITT, which includes people who will not use the tool.

Note that we can also infer the treatment effect from the ITT using a methodology called **instrumental variables**. We will discuss this subject in the next chapter.

A final central aspect of a randomized experiment is the ethical question.

Ethical considerations

Before running an experiment, the implications should be carefully studied. In some cases, showing different versions of an ad might simply affect your sales, whereas in others, it could affect the health of the participants.

Fortunately, research experiments are tightly regulated by various authorities. For example, animal research must follow the *three Rs* principle: **replacement, reduction, and refinement**. If possible, researchers should *replace* animals with alternatives such as cells, tissues, or computational simulations. Additionally, researchers should *reduce* the sample size to the necessary minimum. Of course, the risk is that if the sample size is too small, the results of the statistical tests weaken, and so it might be impossible to test a statistical hypothesis reliably. Hence, statistics allow us to do the sample size calculation properly, providing estimations of the minimal sample size necessary to test a hypothesis. Finally, the third principle, *refinement*, refers to the need to minimize the pain and suffering of the animals. This last principle is achieved by having strict standards of care from the way they are housed to the way treatment is allocated.

These ethical considerations are important beyond medical research. Today, randomized experiments are used in numerous fields, including economics, where Nobel Laureates such as Esther Duflo and Abhijit Banerjee have pioneered their use to study how to alleviate poverty.

A powerful example comes from research on health in Africa, which studied the best way to distribute insecticide-treated bed nets to protect families from malaria (see *Free Distribution or*

Cost-Sharing? Evidence from a Randomized Malaria Prevention Experiment, `https://academic.oup.com/qje/article-abstract/125/1/1/1880305`). To find out what pricing strategy works best, researchers conducted a field experiment in Kenya. They randomly assigned different prices for the bed nets to various prenatal clinics; some clinics offered the nets for free, while others sold them at heavily subsidized prices.

This design immediately raises an ethical question: Is it fair that some pregnant women have to pay for a life-saving product while others get it for free? This is a central challenge in such experiments. In this case, the ethical trade-off was justified because the results were essential for shaping a national health policy that could save millions of lives. The goal was to find the optimal subsidy level to maximize the number of families protected from malaria.

A common assumption might be that charging a small fee would ensure that only those who truly intend to use the nets would acquire them, making the program more cost-effective. However, the experiment revealed the opposite. The study found that demand dropped dramatically when even a small price was introduced. Free distribution led to far higher adoption rates and, according to the researchers' cost-effectiveness analysis, would likely save more lives at a lower cost per life saved compared to cost-sharing programs.

The benefit of answering this question with strong causal evidence was immense. By running this experiment, the researchers provided policymakers with a clear, data-driven answer: if the goal is to maximize health coverage and save the most lives from malaria, distributing the nets for free is the most effective and cost-effective strategy. This demonstrates how the short-term ethical complexities of an experiment can be justified by the long-term benefit of creating optimal, life-saving public policy.

This short-term versus long-term effect plays a central role when considering randomized experiments, often one of the main weaknesses of such methods. So now, let's turn to the limitations of randomized experiments.

Assessing randomized experiments' weaknesses

Randomized experiments are called the gold standard because they can solve numerous issues and help to approach the parallel world situation. But there is no silver bullet; climbing from the valley of despair and escaping the fundamental problem of causal inference is not simple. Actually, we have already seen that it is not "that" simple to run such experiments. Numerous issues can arise at different stages. Randomized experiments allow us to measure a causal effect only if they are made properly and address all the issues mentioned in this chapter.

Even a properly designed experiment is not immune to challenges, such as the tension between short-term and long-term effects I just mentioned.

Let us see how, in the following sections.

External validity in randomized experiments

In randomized experiments, internal validity (did the experiment work correctly?) is a necessary precondition, but external validity (will the results hold in the real world?) is often what we care about most. For example, a drug tested in a perfectly controlled clinical trial must still prove effective for real patients who may take it inconsistently, combine it with other medications, or have different lifestyle behaviors.

These concerns are often lower in business settings. For example, the countless experiments made on websites, online platforms, or apps have a high external validity. In these cases, you are able to observe the behavior of your actual clients while they interact with your product without even knowing that they are part of an experiment.

However, external validity problems with A/B testing in the business world does arise from the temporal comparison. The results can be influenced by seasonality, business cycles, or a competitor's actions. Let me illustrate how results can be misinterpreted. Imagine you compare two ad options, *A* and *B*, in January and find an average 5% **click-through rate** (**CTR**). Then, in August, you test two new options, *C* and *D*, which get a 2% CTR on average. Concluding that *A* and *B* are superior is a mistake. You cannot compare the two experiments because the context (the time of year) is completely different. The lower CTR in August could be due to seasonality, a recent change in your brand's image, or other factors.

Moreover, controlling the environment perfectly in an experiment comes at a cost. Often, such experiments lose all their meaning, and the value for real-world application is low (in other words, the external validity is poor). Imagine that you want to study the question we discussed at the beginning of this chapter: the effect of being further from the teacher in a classroom on the capacity to learn and your test score. Obviously, if you just observe the distance to the professor and grade, the negative correlation could be just explained by the fact that those less interested are seated further from the professor. One way to randomize and approach the parallel world conditions is to use virtual reality (see, for example, *Do students learn better when seated close to the teacher? A virtual classroom study considering individual levels of inattention and hyperactivity-impulsivity*, `https://www.sciencedirect.com/science/article/pii/S0959475217305716`). You could present the exact same class with the same content, same disturbance to different students, and randomly place them in the room.

While this situation sounds ideal, there is a clear flaw: most of the time, we do not live in a virtual classroom. Important factors might be omitted in the virtual classroom. For example, you might be even less focused in reality when you sit at the back because you talk with your neighbor, you play games on your phone, or you answer emails on your laptop. It is important to mention that by using the concepts from our framework, this would create an attenuation bias. All these real-world factors are absent from the virtual classroom. Therefore, the experiment has high internal validity but poor external validity. The causal effect measured in the sterile VR environment, perhaps a small one, might significantly underestimate the true negative impact of sitting at the back in a noisy, real-world classroom, where distractions are a major additional factor.

The problem of reduced external validity in randomized experiments was perfectly illustrated in a peer-reviewed publication in one of the best medical research journals in the world: the *British Medical Journal*. Using a randomized controlled trial, a team of researchers studied the effect of jumping from an airplane with or without a parachute on the major traumatic injuries and cause of death (see *Parachute use to prevent death and major trauma when jumping from aircraft: randomized controlled trial*, `https://www.bmj.com/content/363/bmj.k5094`). Between 2017 and 2018, 23 aircraft passengers participated in the study. Surprisingly, the researchers found no difference between the risk of death or major traumatic injury between the two groups. How is it possible? Well, the participants jumped from a small stationary airplane, on the ground (a height of about 0.6 meters, or 2 feet). This famous research was conveyed and published to raise awareness in the scientific community about this key problem. Some research, including in medicine, will control the environment so well to reduce any noise, disturbance, and differences between groups that it might have poor external validity. This is a common discussion with respect to the use of lab animals, particularly inbred strains of rats or mice, which are very similar genetically and live in the same environment, reducing the capacity to extrapolate to humans, for example.

This tension between internal and external validity also relates to the issue of short-term and long-term comparisons, which I mentioned earlier. In particular, in the digital world (e.g., apps, websites, and social media), users have repeated interactions with different platforms. Hence, it is important to consider the potential difference between short-term and long-term effects. Two opposite forces create issues if the analysis relies only on short-term comparisons.

The first issue is called the **newness effect**. The newness effect (sometimes called **novelty**) refers to the fact that people might click or just try to experience something new. Hence, in the short run, you might witness a lot of clicks on a new button, but once the users experience it, they might stop using it if they don't like this new functionality or this change. This always makes me think about LinkedIn.

Recently, LinkedIn has been pushing for vertical short videos similar to the Instagram Reels or TikTok format. Initially, I was drawn to this feature and spent some time exploring the videos. However, I quickly concluded that I disliked it. The nature of the content and the format didn't align with why I use the platform; when I'm on LinkedIn, I prefer to read more focused, professional content. So, after a short-term increase in my time spent on the platform to discover this new feature, the long-term effect was that it actually drove me away more quickly.

A second similar issue is called the **primacy effect**. Primacy is when experienced users are puzzled by a change, and they take time to adjust. For example, they might momentarily take more time to find the functionality, and then, in the longer term, either use it more or less.

In both cases, it is essential to consider how users' behavior changes over time. Hence, having variability in the sample (short-term and slightly longer-term data) allows you to explore these differences. Variability is also important beyond the time dimension, but with the nature of the different observations. If the sample is highly homogenous, it limits the capacity to extrapolate to everyone. While I do not like the short videos on LinkedIn (at least for the moment), others might like them, as reflected by the high number of reactions to some videos. Hence, studying the heterogeneity of the effect on different groups of users is also central to avoiding relying on an average effect that hides detrimental effects for a subpart of the population. But of course, this requires observing the heterogeneity by having a complete sample of varied individuals.

This is a typical limitation of randomized experiments. Luckily, this is a problem that can be addressed by new causal AI methods, as we discussed in *Chapter 2*. In particular, causal forests allow us to identify, in a data-driven way, the specific subgroups that react differently to a treatment (see *Science of price experimentation at Amazon*, `https://assets.amazon.science/ba/f5/f761c2a04652a798704b5208cc60/science-of-price-experimentation-at-amazon.pdf`).

Why statistics are essential for A/B testing

As A/B testing becomes more popular and accessible, new problems emerge. I met people at companies who would tell me that they do A/B testing. However, while going a bit deeper, I realized that they just clicked on some functionality in an app and compared the results between two options naively and without statistics. As anyone can run an A/B testing campaign using, for example, the toolbox for ads on Meta, this gives the issue of accessibility a false sense of expertise.

The first problem is that many people will be clueless about most of the concepts we discussed in this chapter: external validity, sampling bias, newness, primacy, and so on. This leads to a naive and heroic usage of these tools, leading to suboptimal decisions by clueless users. The worst part is that they don't even know that there is an issue.

This issue goes even deeper. Looking at the results and making any conclusion without statistics can be misleading. When running an A/B test, it is likely that one of the options will be worse than the other. You might be willing to quickly shut down the experiment and only present the most efficient option to avoid negatively affecting your users or clients. This is where statistics come into play. Statistical methods will allow you to assess when you can conclude reliably that there is a difference and that it is not just by chance. To do so, you need to have a sample that is large enough. This is often a limiting factor, as collecting data might be costly or difficult. For example, if your online platform has three users per day, you might never have a sample large enough to test your results in a statistically significant way. To assess the minimal sample size, statisticians can do sample size computation, allowing for the estimation of the minimal sample size before running an experiment.

Moreover, statistical methods exist to minimize the negative effect of the worst options. A method called **multi-armed bandit** allows for quickly adjusting the proportion of the subjects who will be exposed to each variant based on the success of the variants. For example, imagine that you have two versions of a thumbnail of a video. Version *A* quickly reveals a higher CTR. The multi-armed bandit algorithm will start showing more and more version *A*. However, it is still ensured that proper statistical methods are respected, allowing for the separation of noise from signal and concluding, with a statistical test, which option was the most efficient one while minimizing the exposure to the "bad" option. Note that such methods are also risky. For example, if the effect from one version takes more time to kick in, you might shut down the best alternative too early (e.g., if subject to newness/primacy).

Randomized experiments are the gold standard. But as we have seen, it is not always easy to set up once we dig a little bit into the risks. Moreover, one of the main problems that I haven't mentioned yet is that sometimes, it is just impossible to use experimentation to answer some questions.

Randomization is not always possible

Randomization is not only sometimes costly and limited by the capacity to obtain a sample or financial resources, but also it is sometimes just plain impossible to do practically or ethically. It is impossible to change the sex or ethnicity of people to study discrimination. An alternative that has been used extensively is relying on application files to study discrimination in hiring (see *Are Emily and Greg More Employable Than Lakisha and Jamal? A Field Experiment on Labor Market Discrimination*: `https://www.aeaweb.org/articles?id=10.1257/0002828042002561`). Researchers would send identical resumes, only changing the name to influence the recipient's thinking that the person is a woman or a man, for example. But this is limited; you could not test this effect for interviews, for example (maybe using generative AI and deep fakes in the near future).

Imagine trying to test a rebranding for your company. You can't rebrand entirely to see whether it works or not, as it would require changing every logo and visual identity component, leaving no control group to observe. Sure, you could test in a controlled environment, using a survey, for example, and asking for an opinion on your identity. However, this is completely different from observing the actual effect on sales, for example.

Let me share one last set of examples from research. I worked on my first publication for seven years, where I studied the effect of weapons export on the probability and intensity of conflict in Africa. Obviously, it is neither desirable nor possible to start randomly sending weapons to countries to see whether violence would increase or not for the sake of research. Another example is the effect of pollution on health. Researchers are not going to expose humans to lethal doses of pollutants in a lab to assess the risk. We might do this with rats, but the fact that we would need to extrapolate to humans weakens the usefulness of the results (the external validity).

Do not despair, the next chapter will present the other families of solutions allowing for measuring causality when randomization is not possible: **quasi-experimental methods**. We are going to see examples of research answering some of the questions mentioned previously: the effect of weapons on wars, and the effect of pollution on respiratory disease and mortality. But first, be sure to review *Case study 1*, where you'll apply these concepts to evaluate a peer-reviewed scientific article.

Case studies

The following case studies will help you take a step ahead in better understanding the topics we have covered in this chapter.

Case study 1: Cold showers and the immune system

As our analysis has revealed so far with the first case study, the relationship between cold showers and health was confounded with numerous factors, and finding a good control group was difficult.

Hence, a solution is to set up a randomized experiment. Cold showers and the eventual health benefit are not only a trend but have been studied scientifically, covered in magazines, and featured in documentaries. So, now let us have a look at a randomized experiment on this topic. If you look on the internet for cold showers and sick leave, you are likely to find numerous references to one study from 2016: *The Effect of Cold Showering on Health and Work: A Randomized Controlled Trial* (`https://pmc.ncbi.nlm.nih.gov/articles/PMC5025014/`).

In this research, more than 3,000 participants in the Netherlands were randomly allocated to different groups. Participants were included in the study only if they had no severe comorbidities and no prior routine of cold showering. The sample was split into four groups. The control group was taking normal showers every day for 90 consecutive days. The three intervention groups were instructed to shower with warm water for as long as preferred, but ending with 30, 60, or 90 seconds of the coldest available water temperature. This initial intervention period was for 30 consecutive days. Following this, for the next 60 days, they were instructed to shower with cold water at their own discretion.

The main result of the study was that the group with cold exposure experienced a 29% reduction in sickness absence compared to the control group (no cold shower). This result was based on self-reported data.

Let us put your understanding of the concepts to the ultimate test: questioning a peer-reviewed scientific article published in a good journal. Use all the concepts in this chapter to question this result. In particular, think about placebo and blinding, about the fact that all participants enrolled voluntarily, and about the fact that all the results were self-reported (sickness, symptoms, and absence).

Suggested answer

The main issue, mentioned at the very end of the article, is that the results could be influenced by a placebo effect. There is no placebo for cold showers—when you take one, you're very aware of it. While the placebo effect could make it harder to identify the true cause of the benefits, it could still support the case for cold showers, as long as no negative health effects are observed. After all, if you feel better—even just thanks to a placebo effect—why not advocate for cold showers?

There is a more serious issue: people like to be liked and tend to do what they believe is expected of them. This is called **desirability bias**, where participants aim to appear positively in the eyes of the experimenters by making the *right* choices. The participants in this study are volunteers, and it is likely that they are aware that cold showers are expected to boost the immune system and reduce sickness absence. As a result, they might go to work even when feeling ill because they believe that is what is expected of them. This can also be described as the **Hawthorne effect**, where individuals change their behavior simply because they know they are being observed. Maybe if they were not observed, they would stay home more easily when feeling tired or sick. This is also supported by the fact that no statistical difference was found between the treatment and control groups in the number of days participants felt ill.

Additionally, since the experimenters are not physically present with the participants, it is unclear what the true compliance rate is. In other words, we cannot be certain whether participants are actually taking cold showers, what temperature qualifies as "cold," or whether they are adhering to the prescribed duration. However, it is very important to use step 3 from the Causal Mindset framework here: what is the direction of the bias? In this situation, non-compliance would lead to an attenuation effect. Hence, if an effect is observed, and if this were the only problem, we could claim that we observe a lower boundary. If anything, we are just conservative with our estimation, and it might actually be even larger. As we saw in *Chapter 5*, we can live with this; it will not entirely destroy the claim.

Addressing these issues is challenging. As mentioned in the chapter, focusing on more objective metrics that are harder to manipulate, such as various physiological measurements, could help. Additionally, using technological tools such as wearable devices that track temperature could help verify the time spent in cold water. While these tools could still potentially be manipulated, it would be significantly more difficult to do so.

Case study 2: Marketing campaign for energy sobriety

As we've discussed, A/B testing is a valuable strategy for marketing. In the case of the Swiss energy sobriety campaign, the situation was urgent, making a negative control (such as running no campaign at all) an undesirable option. A better approach would be to use a positive control, for example, by A/B testing two different campaign messages against each other to see which one encourages more people to click a link (CTR). If you ran this experiment, what would you be measuring precisely, and what important question would remain unanswered?

Suggested answer

You could test the impact by measuring the probability that people click on the link (CTR). However, you would not be able to directly test the effect of actions taken at home to reduce energy consumption. The next chapter will explore how to measure the behavioral impact, as we will need other methods.

Case study 3: Air pollution and respiratory disease

We have been discussing how to measure the impact of air pollution on health since the beginning of the book with this third case study. We have discussed why comparing respiratory disease prevalence between highly polluted areas with less polluted places was not a valid approach. Randomized experiments might provide some possibilities to answer this question causally, but might also have limitations.

Considering the tools we've discussed in this chapter, what are the possibilities and, more importantly, the ethical limitations of using randomized experiments to study the effect of air pollution on respiratory disease?

Suggested answer

Randomized experiments offer important answers but leave some aspects open. We could, for example, conduct randomized experiments on animals, provided the study is well-justified and complies with strict ethical standards. Such an experiment would allow us to quantify the causal effect between specific pollutants and respiratory effects in a highly controlled environment.

However, while we might be able to justify this on lab animals, we would not expose humans to a lethal dose of pollutants. And a central question is the effect on humans. In this situation, we can clearly discuss external validity and, in particular, the capacity to extrapolate to humans. This is a classic example where the "gold standard" of randomization is impossible to apply directly to the population we care about. Luckily, the other family of methods presented in the next chapter will allow us to fill this gap.

Summary

In this chapter, we began our steep climb up the slope of enlightenment, leaving the "valley of despair" behind. You learned about the so-called "gold standard" for measuring a causal effect: the randomized experiment. We saw how this powerful approach, known as randomized controlled trials in research and often referred to as A/B testing in business, is our best attempt at recreating the ideal "parallel worlds" scenario. You learned the critical steps of a good experiment, from choosing a representative sample to the crucial role of randomization in breaking the link with confounders, and the subtle but important details of blinding and placebos.

But, as we discovered, there is no silver bullet. The gold standard has its limits. We ended by seeing that perfect experiments are often practically impossible (we can't randomize a person's ethnicity to study discrimination during interviews) or simply unethical (we can't randomly start wars to measure the effect of weapons or expose humans to a lethal dose of pollutants). Luckily, there is an entire family of methods allowing us to address these limitations, which we are going to explore in the next chapter.

Join our community on Discord

Join our community's Discord space for discussions with the authors and other readers: `https://discord.gg/G4YmB6DQvf`

7

Quasi-Experimental Methods

Equipped with our tools and with the first solutions to assess causality, we are now already far from the valley of despair. Yet, the road to causality is not straight, nor simple. At the end of the last chapter, we took a small step back when we saw the limitations of randomized experiments: they are often impossible or unethical to carry out.

This is exactly why studying causal inference matters. The quest for causality is never trivial; it is full of obstacles. But that is also what makes it rewarding and useful. We haven't said our last words. There is another set of methods, complementary to randomized experiments, that often helps us overcome these limitations: **quasi-experimental methods**.

Quasi-experimental methods allow us to infer causality using observational data when the treatment is not randomly assigned. In this chapter, we will study the key concepts and the most popular methodologies:

- Understanding identifying assumptions
- Difference-in-differences
- Regression discontinuity design
- Instrumental variables

Understanding identifying assumptions

Measuring a causal effect relies on the capacity to find a good reference point (proxy counterfactual), allowing us to get closer to the parallel world situation and reducing the risk of confounders. The core idea is that we find a situation where the control and treatment groups have just one systematic difference affecting the outcome, which is the treatment.

In a randomized experiment, proper randomization and design allow this. However, with the quasi-experimental method, the experimenter can't fully randomize and control the treatment allocation. Hence, making a causal claim will require some specific assumptions.

The central piece is the identifying assumption: the condition that must hold in order for us to interpret the estimated effect as causal. But there is a catch. Generally speaking, you can't test this assumption directly. This is a direct consequence of the fundamental problem of causal inference, the absence of parallel worlds, as we discussed in *Chapter 3*.

In essence, the identifying assumption refers to the fact that the treatment is the only systematic difference affecting the outcome between the treated and control group. However, this assumption will take a specific form, depending on the method used. Moreover, while we can't directly test this assumption, we can provide support for its validity with other indirect tests.

You understood it; questioning the identifying assumption is central to challenging or supporting a causal claim. This will require the use of domain and technical knowledge. I will discuss the identifying assumption for each method in the following section.

Let's start with the first method.

Difference-in-differences

To illustrate the challenges and idea of conducting a quasi-experimental study with our first method, let's continue where we stopped in the last chapter: How can we assess the impact of pollution on respiratory diseases? As we have seen, while it is possible to use animals in a lab to study this issue, it would be unethical to expose humans to dangerous levels of pollution. Even if animal research is tremendously useful, it can be difficult to generalize the findings to humans, who may be exposed to pollution over a longer period of time, particularly in the complex context of their daily lives.

As we have already discussed, beyond research on animals in a lab, while we could compare people living in highly polluted areas to those living in less polluted areas, this comparison would be flawed due to the many potential confounders (such as their diet, physical activity, and type of work).

Hence, to answer such questions, we can resort to quasi-experimental methods. That is exactly what some researchers did in 2016 (see *The effect of air pollution on mortality in China: Evidence from the 2008 Beijing Olympic Games*, `https://www.sciencedirect.com/science/article/abs/pii/S0095069616300237`). The authors used the Olympic Games in Beijing to measure the causal effect of air pollution on mortality. The government imposed very strict laws to reduce pollution in the city before and during the Games (e.g., the closure of power plants and a reduction in the use of cars).

Technical comment

In the paper, the author used a more complex approach than a simple difference in differences. However, this detail goes beyond the scope of this book and won't prevent us from using this paper as an illustrative example while ignoring this technicality.

The starting point of any causal inference endeavor is to find a good counterfactual, a place similar to the treated place but without the treatment. Normally, comparing places with higher and lower pollution isn't valid because of confounders. But could the Beijing Olympics give us a special case where this works? Here, we would compare the same place but just at different points in time. But wait a minute, what if we ask, *Is there something else?*. Yes, there is an important confounder. The flaw with a simple before-and-after design is that it conflates seasonality with the effect of the policy. While we fix the spatial dimension (looking at the same place before and after), the time dimension is different. Air pollution in Beijing is influenced by weather and seasonality. The Olympic Games took place in summer, a period when pollution is often slightly lower anyway in Beijing. This creates an amplification bias: we expect the air pollution policies to reduce pollution, but part of the observed drop may simply be due to the season. Many other things might be subject to seasonality, from pollution to social events. Imagine that people tend to smoke less when it is colder, as they have to stay inside. Or maybe they do more sports during the summer to prepare their "summer body." Long story short, there are several things changing over time affecting health and hence confounding the potential effect of the air pollution policies.

Using our Causal Mindset, we managed to invalidate this before-and-after comparison. On the one hand, comparing different places during the Olympic Games is not valid because there might be several systematic differences between the two groups. On the other hand, focusing on Beijing and doing a simple before-and-after comparison is also flawed.

Each of the two strategies, comparing the same place over time or comparing different places at the same time, has some advantages, but also serious flaws. What if we could combine the two types of comparisons to benefit from the pros while addressing the limitations? This is exactly the idea of one of the most popular quasi-experimental methods: difference-in-differences.

The idea of the difference-in-differences (or **diff-in-diff**) is to compare the evolution of the outcome through time between a treated and a control group. The key is to find two places that evolve similarly before the treatment period and then compare the evolution between the two during the treatment period. This idea is represented in the following graph:

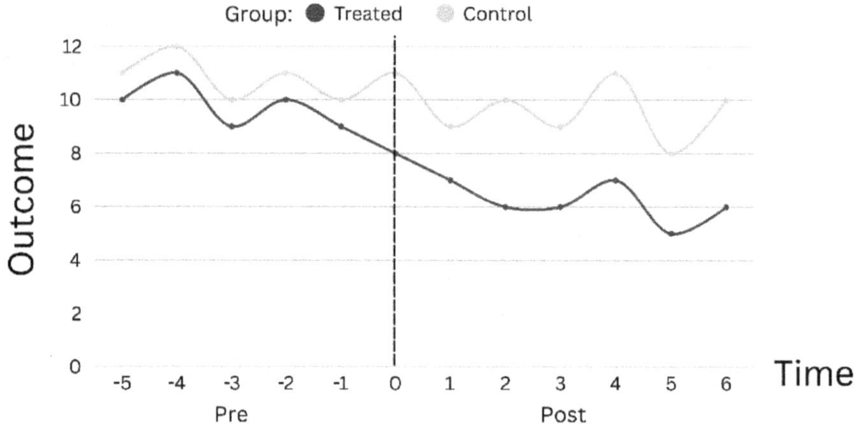

Figure 7.1: Visual representation of the canonical difference-in-differences setup. The vertical dashed line represents the beginning of the treatment period

In the context of the Beijing Olympic Games, researchers used this methodology to assess the causal impact of the air pollution policies on mortality. Difference-in-differences is called as such, as it is computed by subtracting two differences. The main computation is represented in the following graph:

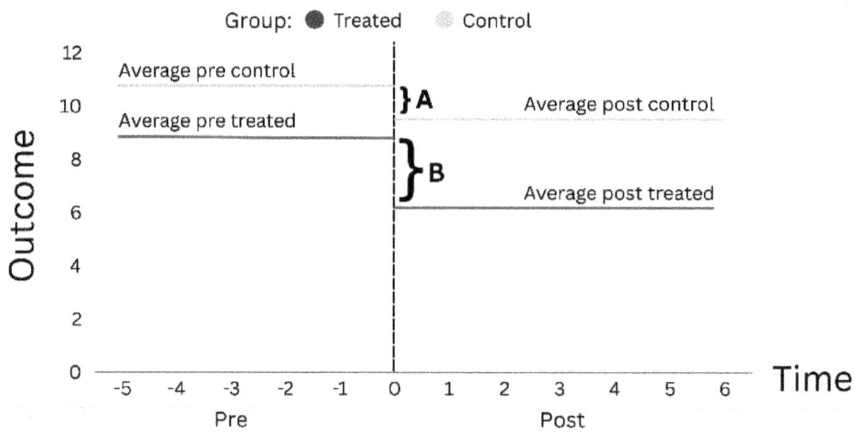

Figure 7.2: Visual representation of the main computation of the difference-in-differences method

The first difference represents the variation between the post- and pre-period for the treated units, denoted as **B** in *Figure 7.2*. This difference represents the variation in outcome for the treated units, which includes the treatment effect confounded with some other effects, for example, seasonality. We also need to compute the difference between the post- and pre-period for the control group, denoted as **A** in the figure. This difference captures the variation through time without the treatment. This could be seasonality, for example. The final computation requires subtracting the two differences: $B - A$.

If we have a good control group, this final subtraction allows us to identify the treatment effect. Let's look at it in a simplified form:

- **A** = *Seasonal effect*
- **B** = *Seasonal effect + Treatment effect*
- **B – A** = (*Seasonal effect + Treatment effect*) – *Seasonal effect = Treatment effect*

We can see that this result holds only if the control group shares the same seasonality as the treatment group; otherwise, the second difference cannot cancel out the confounder.

Using this methodology, the research cited previously arguably measured the causal effect of the air pollution policies around the Beijing Olympics to infer the benefit of such policies, but also informed us on the lethal effect of air pollution for humans. The researchers compared the mortality between the regions affected by the policies and the control regions not affected by the policies. This approach fills the gap left by randomized experiments. The main result is the following: "*We find that a 10 percent decrease in PM10 [(small particulate matter in the air)] concentrations reduces the monthly standardized all-cause mortality rate by 8 percent.*" In the paper, they separate by cause of death and reveal that these results are driven by air pollution-related deaths, defined as those caused by **cardio-cerebrovascular and respiratory** (**CVR**) diseases, while no effect is found on air pollution-irrelevant deaths (e.g., car crashes).

As usual, the core idea is to find a good counterfactual. In the diff-in-diffs setup, that means identifying another group of observations that followed a similar path in the past. The period before treatment starts is called the **pre-period**. In *Figure 7.1*, we see both groups moving along the same path during this pre-period.

Choosing a good control group requires a mix of domain knowledge and technical insight. Domain knowledge narrows down which groups are even plausible proxy counterfactuals; you need a solid rationale for why another set of observations might serve as a valid comparison.

For example, in the Beijing Olympics study mentioned earlier, the authors compared regions affected by the pollution policies to other regions in China that were not regulated. Here, domain expertise guided the choice of control units by focusing on regions with similar characteristics and behavior. After that, technical tools come into play to formally assess whether the identifying assumption holds in this setup.

Identifying assumption for difference in differences

In the diff-in-diff setting, the key identifying assumption is called the **parallel trends assumption**. This assumption means that, without the treatment, the treated and control groups would have experienced the same change in outcomes over time, that is, their trends would have been parallel.

This is impossible to test, as we do not observe the treated group in the treatment period without the treatment. However, we do observe the two groups before the treatment starts, and hence, we can test whether the two groups follow a parallel path in the pre-period. If statistical analysis of the data does not reveal differences in trends in the pre-period, this gives us some confidence that the identifying assumption is reasonable.

Hence, this is one of the first checks you should perform before estimating the causal effect in a difference-in-differences setting. If the two groups do not follow a parallel path in the pre-period, it reveals that the untreated group is not a valid proxy counterfactual, because without the treatment the two groups do not move in a similar way. If you want to go deeper into the technical aspects and statistics behind this methodology and test, I recommend you read the fantastic paper *Difference-in-Differences Designs: A Practitioner's Guide* (`https://arxiv.org/html/2503.13323v1`).

But statistical checks alone are never sufficient. Because we never observe the treated group in the post-period without treatment, we must also rely on domain knowledge to judge whether other shocks may have affected only one group during that time. Combining both statistical evidence and substantive reasoning is essential to make a credible causal claim.

Imagine that in our example of air pollution, new coal power plants have been built in these control regions in order to answer the growing energy demand in China, or imagine that an outbreak of a serious respiratory disease occurred, but it was better controlled in Beijing to protect Olympic participants. In both cases, another systematic difference between the groups would also influence health outcomes, violating the identifying assumption. Notice that I emphasize differences that affect the outcome. Only those count as confounders. Random differences that do not influence the outcome would not threaten causality.

Essentially, questioning the parallel trend assumption is a complex way of asking, *Is there something else?*. Imagine that you want to use this method to evaluate the effect of a new policy, for example, a reform of unemployment benefits. You compare country A, which implemented this reform, with another similar country, B, which didn't change its unemployment system. However, imagine that country A made this change due to an economic crisis in the country. This crisis would create another major difference between the two countries that might also affect the unemployment growth rate in country A. Hence, the effect of the economic crisis would be confounded with the effect of the reform. Depending on the assumptions and situation, this example could also illustrate a problem of reverse causation.

Now that we've seen the logic and potential pitfalls of the method, let's look at concrete examples to see how the diff-in-diff approach is applied.

Difference-in-differences examples

This popular diff-in-diffs method has been used in several papers recently to estimate the effect of generative AI on employment. In *Who Is AI Replacing? The Impact of Generative AI on Online Freelancing Platforms* (`https://pubsonline.informs.org/doi/10.1287/mnsc.2024.05420`), three researchers studied the impact of ChatGPT 3.5's release in November 2022, as well as other generative AI tools for image generation on a major global freelancing platform. The advantage of studying freelancing is that it is a fast-moving market, allowing us to quickly observe the impact of the shock. The researchers compared the evolution of job postings for jobs that are more exposed to generative AI (e.g., writing or web development) compared to other jobs less exposed to automation with generative AI (e.g., more manually intensive jobs such as office management or audio service). The author found "*a 21% decrease in the number of job posts for automation-prone jobs related to writing and coding, compared to jobs requiring manual-intensive skills, within eight months after the introduction of ChatGPT.*"

The same methodology is often used in the business world when you can't randomize the treatment allocation. Moreover, the diff-in-diff approach allows for the measurement of aggregated, often longer-term effects (over several weeks, months, or years), which is beneficial in scenarios where we expect a slow-moving process or a lasting impact.

It could be used, for example, to measure the impact of a traditional TV ad campaign on online sales. Imagine you want to assess the return on investment of a new TV ad campaign that runs in certain cities but not others. Here, you have no control over which individuals see the TV ad, so a classic randomized experiment is not possible. In this case, the best you can do is look for a similar city where the ads were not shown to act as a counterfactual. Then, by comparing the change in online sales over time in the cities with the ad campaign (the treatment) to the change in the control cities, and assuming that no other systematic differences exist between them, you can approach a causal effect with the diff-in-diff approach. Matheus Facure also illustrates the diff-in-diff method to measure the impact of a billboard campaign in his book *Causal Inference for the Brave and True*.

Let me add another example from the retail world. A crucial question for any local business is: what really happens to my sales when a giant such as Wal-Mart opens a new store nearby? Simply comparing stores in towns with a Wal-Mart to those without one is a flawed approach. As you can guess, Wal-Mart doesn't choose its locations randomly. It strategically enters markets that are promising. This creates a classic confounding problem. To solve this, researchers used a diff-in-diff strategy (see *When Wal-Mart Enters: How Incumbent Retailers React and how this Affects their Sales Outcomes*, `https://journals.sagepub.com/doi/10.1509/jmkr.47.4.577`). They identified a set of stores where a Wal-Mart opened (the "treatment" group) and compared their sales trend to a carefully chosen control group: other stores from the same retail chain in similar markets where a Wal-Mart did not open. By comparing the change in sales before and after the opening for both groups, they could isolate the true causal impact of their new, massive competitor.

As we've seen, the diff-in-diff method requires time-varying data across multiple units. While the allocation between treated and control groups is rarely random, the ability to observe both a before and after period allows us to identify causal effects, provided that the parallel trends assumption holds. However, this assumption can sometimes be difficult to defend.

The next method we will explore, the regression discontinuity design, does not require groups to evolve similarly over time or even time-varying data. Instead, it takes advantage of situations where an arbitrary rule or cutoff generates a quasi-random assignment into treatment and control groups.

Regression discontinuity design

The idea behind a **regression discontinuity design** (**RDD**) is to exploit discontinuity in the treatment allocation (e.g., a geographical border, a certain score on a test, etc.). If you focus on observations near the discontinuity, you can observe very similar people or places to either receive the treatment or not.

In nature, most of the boundaries are smooth. For example, if you want to compare people living where it is very hot with people living where it is very cold, they are far away geographically, and hence many differences exist between the two groups other than just the temperature. Similarly, if you compare young people with older people, age is different, but also their daily activities and health risks. However, humans like to create rules based on thresholds. A state border is a perfect example of this, creating spatial discontinuity. On one side of the border, a particular social policy might exist, while on the other side, it might not. Suddenly, two groups of people who are geographically neighbors receive different treatment, which is eventually the only systematic difference. If this is true, the absence of confounders would allow us to pin down the causal effect. In causal inference, we often say the allocation is *as good as random*: individuals just above and just below the cutoff are nearly identical, except for whether they receive the treatment.

In this context, the identifying assumption is that nothing else systematically changes at the threshold. If that holds, we can attribute the observed jump entirely to the treatment.

Let me illustrate with a concrete example. Consider scholarships for students from modest financial backgrounds. Imagine you want to measure the impact of such scholarships on wages later in life. To do so, you compare the wages of the students who received the scholarship with those who didn't. There is a clear confounder: parents' income. Those who didn't receive the grant (and probably didn't ask for it) are also those with higher financial resources. Higher financial resources might allow the students to support their learning (e.g., paying for a tutor, a personal computer, books, etc.). This situation is illustrated in the following figure with a directed graph:

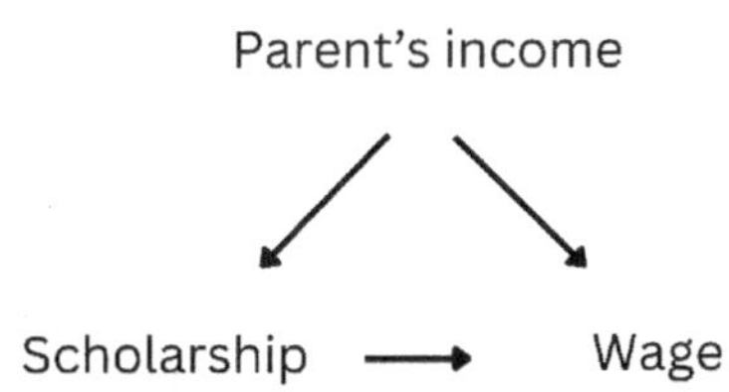

Figure 7.3: Direct graph representing the impact of scholarship on wage later in life

Simply comparing the average wages later in life of students who received a scholarship with those who did not would be biased, because of confounders. A better approach is to compare students right around the eligibility cutoff, for example, those just below and just above the income threshold.

Suppose the threshold is a parental income of \$50,000 per year. A family earning \$49,900 and a family earning \$50,100 have essentially the same financial resources. The small \$200 difference is unlikely to have any real impact on a child's future wages. But in this case, crossing the threshold determines eligibility: one child receives the scholarship, the other does not. As long as there are no other shocks (sudden changes) around the threshold, we can attribute this jump (discontinuity) to the cause (the scholarship).

This setup is almost like a random assignment. The scholarship is the only systematic difference between the two groups, and any later differences in wages can therefore be attributed to it. This approach is called the regression discontinuity design, as it exploits a discontinuity in the assignment rule to identify a causal effect.

The following figure depicts the potential effect of the scholarship:

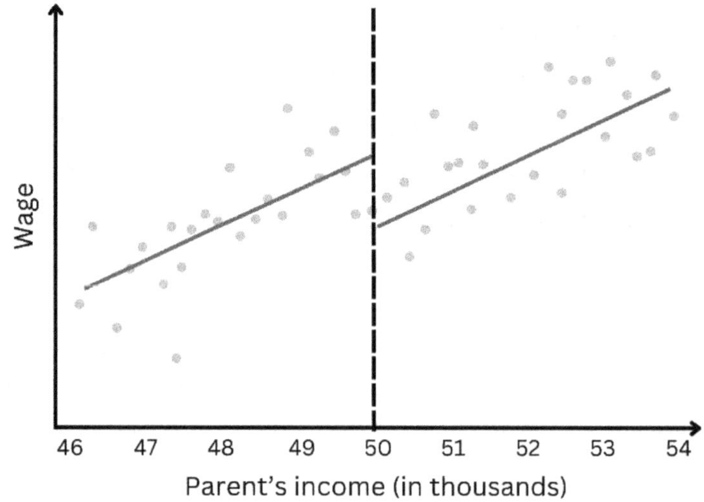

Figure 7.4: Example of a regression discontinuity

Figure 7.4 perfectly illustrates the idea behind the RDD. The wage of the parent might capture better financial resources, allowing them to invest more in the education of the children, for example. This is well captured by the visible positive trends. In addition, there is a sudden drop at the $50,000 mark. This is our discontinuity and what will be used to estimate the causal impact of the scholarship.

While this sounds very close to a random allocation, we should always be careful and question the identifying assumption.

Identifying assumption for RDD

The central idea of RDD is to exploit a discontinuity in treatment assignment, which can then generate a corresponding discontinuity in the outcome. In other words, RDD lets us isolate causal effects when treatment is determined by a cutoff, and we assume units just above and below the threshold do not have any other systematic difference.

The running variable, the variable that determines treatment assignment based on whether it crosses a cutoff, usually relates to the outcome. For example, in *Figure 7.4*, we can clearly see a positive relationship between parents' income (the running variable) and their child's future wage (the outcome). But our goal is not to estimate the overall effect of parental income on wages, since that would also capture many other factors unrelated to the scholarship. Instead, the idea is to exploit the discontinuity at the allocation threshold: in this example, children from families earning just under $50,000 per year are eligible for the scholarship, while those just above are not.

In this context, the identifying assumption is that there are no other systematic differences around the threshold. If that holds, we can attribute the observed jump entirely to the cause of interest, in this case, scholarship access. Put simply, it comes back to our usual question: *Is there something else explaining the effect?*. As we have seen earlier we often say in this case that the allocation is *as good as random*; as individuals just above and just below the cutoff are nearly identical, except for their treatment status.

Violations of the identifying assumption can arise for several reasons. Domain knowledge is central here to identify these potential issues. The first reason is that there are other discontinuities around this threshold—for example, if another policy uses a similar threshold (e.g., subsidies for health insurance).

Second, it is important to consider whether subjects can manipulate the treatment assignment. Basically, by knowing the value of the threshold, people can change their behavior to end up on one side or the other of the threshold. For example, imagine that in the current example, some people manage to reduce their income slightly to end up below the $50,000 threshold in order to receive the scholarship. If this is the case, the group of people right below and right above the threshold would differ in other characteristics systematically. There would be another discontinuity. Right below the threshold, you would have an over-representation of people who really want their children to continue with their studies, as they put some effort into changing their income in order to get the scholarship. This difference in motivation might also affect the outcome and would therefore be confounded with the outcome.

How to validate the identifying assumption

While we can't directly prove the identifying assumption, we can use two different tests to support it.

Covariate continuity

First, we check the continuity of other variables. Do observable characteristics—parents' education, family size, or pre-treatment outcomes such as high-school IQ—also jump at the $50,000 threshold? If they're smooth, it supports the idea that only scholarship eligibility changes at the cutoff. If we see a jump, something else is going on.

Manipulation risk

A discontinuity in these covariates can indicate manipulation of the running variable: just below the threshold, you may have an over-representation of highly motivated or better-informed families who adjusted reported income to qualify. That motivation can affect later wages and would bias the estimated effect.

Placebo cutoffs

Second, we can run placebo tests (also called falsification tests). For instance, we might test for a jump in the outcome at an arbitrary income level where no policy changes such as $45,000. If we find a jump there, it's a red flag. But if we find no jump at any of the "placebo" thresholds, it increases our confidence that the jump at the true threshold is the real deal.

Focus near the threshold

To further support the identifying assumption, it is important to focus on individuals very close to the threshold. The farther we move away, the greater the risk that people on each side differ in other characteristics. However, the closer to the threshold, the smaller the sample, and hence, it might be harder to have enough evidence to test our hypothesis statistically. Moreover, as the ideal distance to the threshold to balance these two forces is not clear-cut, we often test with several distances to the threshold. Slightly varying the inclusion criteria (distance to the threshold) should not affect the result significantly. If these small tweaks completely change the conclusions, it is a sign that the results are not robust and that the causal claim should be treated with caution.

Limitation: local treatment effects

But this comes at a cost. Narrowing the sample around the cutoff increases internal validity, yet reduces external validity, because the individuals studied become more and more similar. This is why we say RDD estimates a *local treatment effect*: an effect valid for those near the threshold, but not necessarily generalizable to individuals far from it.

Now that we have seen the challenges of identification, we can explore more academic and business examples.

RDD examples

The first example I would like to share is a paper I have discovered in Matheus Facure's e-book: *Causal Inference for the Brave and True*. In this paper, *The Effect of Alcohol Consumption on Mortality* (Carpenter and Dobkin, 2009, `https://pmc.ncbi.nlm.nih.gov/articles/PMC2846371/`), the authors used regression discontinuity to study the immediate effect of alcohol consumption on mortality. The mortality rate related to moving vehicle accidents jumps suddenly when people turn 21 years old, arguably caused by alcohol consumption.

The paper's core idea is that people who are allowed to legally drink alcohol (e.g., those 21 years or older) are typically not comparable to those who are not legally allowed (e.g., younger individuals) due to various factors correlated with age and mortality, such as health risks and behaviors. Additionally, comparing adults who consume alcohol with those who don't can introduce other confounding differences, including psychological health, cultural factors, and socio-economic status. However, by comparing individuals who are just about to turn 21 with those who have recently turned 21, the study compares two groups that are very similar in most respects. The critical condition for this comparison, the identifying assumption, is that there are no other major changes or discontinuities occurring at the same age (apart from the legal drinking age), ensuring that the observed differences can be attributed to the effect of legal alcohol access. Of course, in this paper, they do several placebo tests to support the validity of this assumption. More specifically, they show that demographic characteristics, employment status, insurance status, and educational attainment all vary smoothly around the threshold.

The following graph presents the main results:

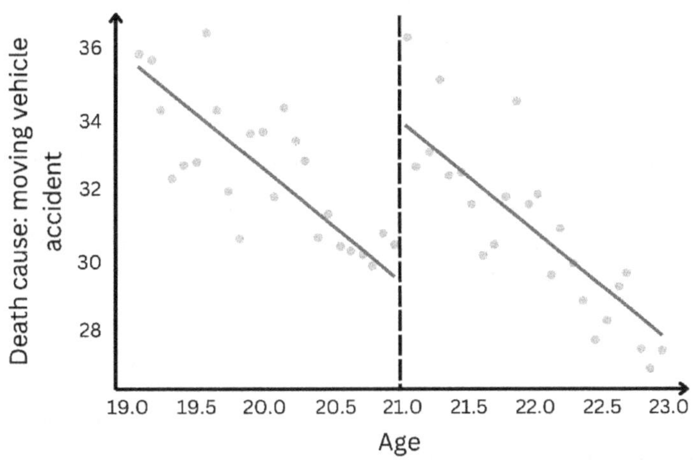

Figure 7.5: The results of an RDD around the legal drinking age on the death rate from motor vehicle accidents per 100,000 people

Many things evolve with age that might be affecting the mortality rate (e.g., varied health issues). However, those risks tend to evolve smoothly, as illustrated in the following graph:

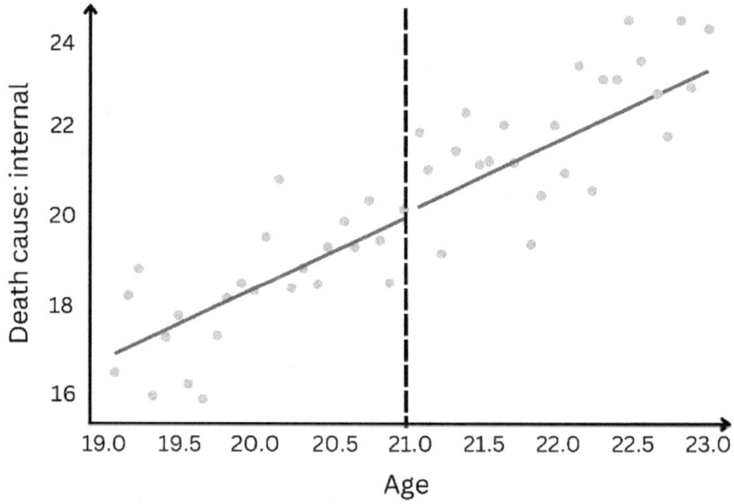

Figure 7.6: The results of an RDD around the legal drinking age on the death rate from internal causes (e.g., diseases)

This second graph represents the exact same model, but instead of predicting road accidents, it aims to predict the death rate from disease. We can see that evolution does not have any clear discontinuity.

Despite these relatively robust results, we can rely on our toolbox to identify some potential weaknesses. First, it is important to keep in mind that RDD measures a local treatment effect. Hence, this strong effect on death by moving vehicle, for example, might be hard to extrapolate to other age groups, since 21-year-olds are relatively young drivers with different experience and behavior on the road than older drivers. Second, the authors note that the effect is largely driven by white males who are high school graduates or attending college. This is a good reminder to go beyond the average and study the heterogeneity of the effect. Moreover, in the paper, they also use reported alcohol consumption for numerous analyses. The authors mention the risk of desirability bias, which might lead people to under-report their consumption, particularly if they are not yet 21 years old, as it is illegal. Overall, this research is an excellent example of causal inquiry. It is rigorous, convincing, and transparent with respect to the potential limitations.

The discontinuity method can also be used in industry. For example, a company might roll out a new training program to employees whose performance scores are just below a threshold and compare them with those just above it. The idea is to learn the effect right around the cutoff, where people are otherwise very similar.

Such methods could also be exploited for marketing or sales. Imagine a situation where a company offers a voucher to consumers who spend more than a certain amount, and you want to evaluate the impact on sales of offering this voucher. By comparing clients with spending close to the threshold, you have people with similar spending levels. However, this example is perfect to highlight an important concept: individuals should not be able to "manipulate" the treatment allocation. If the threshold is known, some consumers might aim to spend just enough to surpass it and obtain the voucher, while others who are less attentive or less financially constrained might ignore the threshold entirely. If this occurs, the people just below and just above the threshold would systematically differ in ways unrelated to the voucher itself, introducing confounding factors. This would make it difficult to measure an unbiased causal effect, as the differences between the two groups could no longer be attributed solely to the voucher.

Some practical safeguards are often used in this specific case. First, check whether many more customers than expected cluster just above the threshold, which would be a sign of manipulation. Second, if needed, drop a very narrow band of observations right at the cutoff and compare slightly farther on each side. Also, remember that in many real settings, you estimate the effect of qualifying for the voucher (eligibility), since not everyone who qualifies actually redeems it. Finally, the result is still local; it applies to customers near the threshold, and you should check that no other policy changes at exactly the same amount.

Another common business question for RDD is: Do promotional emails actually make customers spend more? A simple comparison is flawed, as businesses select who to send promotions based on their behavior, creating selection bias. This question was studied in *Identifying Causal Marketing Mix Effects Using a Regression Discontinuity Design* (`https://pubsonline.informs.org/doi/abs/10.1287/mksc.1110.0670`). To solve this, researchers studied a casino that used a "predicted profitability" score to target its customers. The casino sent a promotional email to everyone with a score above a specific threshold, but not to anyone just below it. By comparing the spending of customers who were virtually identical but happened to fall on opposite sides of this sharp, arbitrary line, the researchers could isolate the true causal effect of receiving the email, free from the confounding effect of pre-existing customer loyalty.

As we've seen, the power of regression discontinuity comes from finding a clean, sharp cutoff in the world, a rule, a score, or a border that creates a natural experiment. But what happens when there is no such convenient rule? What happens in the case of reverse causation?

For these other situations, we need a different kind of tool. We need a way to find an external event, a kind of *nudge* that encourages some people to take the treatment but not others, without directly affecting their outcomes. This brings us to the ingenious and powerful world of instrumental variables.

Instrumental variables

Continuing our quest to understand quasi-experimental methods, we now turn to a complex topic I worked on for seven years: *Weapons and war: The effect of arms transfers on internal conflict* (`https://www.sciencedirect.com/science/article/pii/S0304387822001432`). This work illustrates again how quasi-experimental methods help to fill the gap when it is impossible or unethical to run a randomized experiment.

While you might think that it is obvious that more weapons lead to more violence, some people might disagree. Typically, arms lobbyists would argue that weapons can be used for protection. Beyond this relatively weak argument, quantifying the causal effect is central for policy-making. To motivate political intervention, a clear and precise assessment of the benefit is needed to justify the enforcement of what could be a costly policy. The precision that causal inference can bring is a necessity for decision-making. If a company invests millions in AI, it needs to know what the return is precisely. Saying that *it works, productivity increased*, is not enough to justify spending more money to deploy and maintain such a tool.

Beyond confounders, the main difficulty in the paper I mentioned comes from the fact that weapons might affect wars, but wars clearly affect weapon imports (reverse causation).

To solve this puzzle, I spent years studying weapons and the arms trade. I read books, researched, discussed with experts in Non-governmental organization or weapons enthusiasts, went to a shooting range to study and shoot with an AK-47 and other weapons, and studied the production process and every single detail. I was building the domain knowledge necessary to find a solution to this problem.

One day, I saw an article in the news titled *Ammo shortage squeezes police forces: Demand skyrockets partially due to Wars* (NBC News, 2007, `https://www.nbcnews.com/id/wbna20322566`). I started digging in this direction and realized that I was onto something. When the U.S. intervened in Iraq and Afghanistan, it sent tons of weapons and ammunition to the Middle East to support the war effort. The shock was so severe that police forces in the U.S. experienced a shortage and could not train properly.

This initial anecdotal evidence turned out to be not so anecdotal, as similar situations were observed across the globe. When a major arms-supplying nation went to war, its own military's needs took priority. Supply and production capacity were redirected inward, reducing the availability of its weapons on the global market.

I exploited these supply shocks—shortages caused by a supplier's involvement in a war—to study the consequences for their usual importing countries.

Let me illustrate with a well-documented case. At the beginning of the 21st century, separatist groups were fighting for the independence of the Cabinda region in Angola. In 2002, Spain, a primary arms provider to Angola, engaged in *Operation Enduring Freedom* against Al-Qaeda in Afghanistan. The same year, Angola's arms imports dropped by nearly 90%. The conflict in the Cabinda region was put to a halt. *Human Rights Watch* has well documented the situation and argued that a major factor for this halt was the low availability of weapons and ammunition for the separatist groups. Note that Spain was involved in a conflict on another continent, not directly related to the strategic situation in the Cabinda region.

Hence, I could generalize this approach and use such historical shocks (involvement of weapons suppliers in wars outside Africa) to study the impact on the violence for their usual customers in Africa.

This strategy is technically called an **instrumental variable**. The aim is to find something that affects the treatment (here, arms imports), but which is related to the outcome (conflict in the destination country) only through the treatment and not directly. Let us now discuss the key conditions for having a valid identification with an instrumental variable.

Identifying assumptions for an instrumental variable

There are two main important conditions to have a valid identification strategy with an instrumental variable. These conditions are represented on the following simple directed graph:

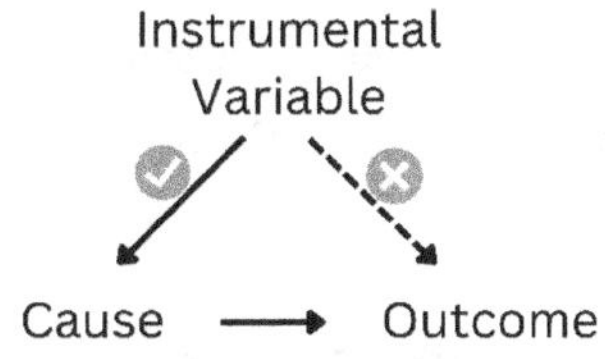

Figure 7.7: Simple directed graph representing the two conditions required to have a valid instrumental variable

The first condition is represented in *Figure 7.7* by the arrow from *Instrumental Variable* to *Cause* with a checkmark. This means that the instrumental variable must influence the factor (the cause) whose effect on the outcome we want to study. In our first example, this means that the supplier's involvement in a war on another continent (my instrumental variable) must influence the weapons imported into the recipient country. This condition can be tested with statistics. This is what we call the relevance condition. Note that I refer here to a condition, as this is not just an assumption; it is something we can test.

Now, let us turn to the second condition. The instrumental variable should not influence the outcome directly, only through the cause. This is called the **exclusion restriction**, and this is the central identifying assumption for an instrumental variable.

You cannot prove directly that this assumption is valid, but there are many things you can do to support its validity. The starting point is to build domain knowledge, which allows you to see why the assumption might be valid and what might violate this assumption. You have to discard all the potential stories that would invalidate this assumption. In this example of weapons and wars, a potential violation came from the fact that some armed groups are active on different continents. Hence, a war in Asia that creates a shortage of weapons in Africa might directly affect a conflict in Africa due to these international connections (and not only through arms shortage). To address this issue, I studied detailed reports on ties between armed groups from various continents, and I excluded all the countries with such ties. Then, I reran my analysis without these countries and found that my results were even stronger, which reinforced the credibility of my findings.

But, of course, this was just one potential violation of my exclusion restriction. For years, I presented this paper to experts from academia to Non-governmental organizations, and people would suggest why this story might be flawed. Every time I would get back to work and use data and statistics, I would try to see whether their point was valid or not. This enriching process led me to study the relationship with the black market, or the role of other imports, or of humanitarian aid, and other economic ties. My strategy survived all those valuable comments and tests and was published in the *Journal of Development Economics*, a top field peer-reviewed journal. The main finding was that weapons imports increase the probability and the intensity of internal conflicts as well as the number of refugees.

Examples of instrumental variables

Let me now turn to a business example of instrumental variables. Finding a good instrument is the hardest part of this method. Several examples are presented and limitations discussed in the following paper: *Even the Rich Can Make Themselves Poor: A Critical Examination of IV Methods in*

Marketing Applications (https://pubsonline.informs.org/doi/10.1287/mksc.2014.0860). A classic marketing example, which this paper discusses, is the challenge of measuring the true effect of price on sales.

A clever and often valid instrument is the wholesale price. The logic is that changes in the manufacturer's costs affect the retail price but don't directly influence a customer's decision at the shelf. However, a very common but often invalid instrument is using last week's price. As illustrated in the paper I referred to, this is a mistake. If a product was on sale last week, customers likely stocked up. Their high household inventory will directly reduce their demand this week, meaning last week's price violates the crucial exclusion restriction.

In practice, instrumental variables are rather complex, as we have seen with the *Weapons and Wars* paper. If you want to go further and study such advanced application in business, there is one interesting post shared by Spotify on their blog, *Encouragement Designs and Instrumental Variables for A/B Testing* (https://engineering.atspotify.com/2023/08/encouragement-designs-and-instrumental-variables-for-a-b-testing). In this article, the Spotify team explains how to measure the impact of a new, exclusive, and potentially very expensive podcast deal on retention, one of their most important goals. Of course, Spotify can't force people to listen to the podcast to measure its impact. Instead, they use encouragement: a randomly shown notification suggesting users listen to the new podcast.

This notification can then be used as an instrumental variable. The notification should increase the probability that a user listens to the podcast (the relevance condition), and because it is shown randomly, it's plausible to assume that its only impact on retention is through the act of listening (the exclusion restriction).

Let's return to our sleeping pill example from the previous chapter. The treatment was randomized; some people were assigned the pill, others a placebo, but maybe some people in the treatment group didn't actually take it. This non-compliance makes us observe only the effect of the intention to treat and not the actual treatment effect.

Instrumental variables help us recover the treatment effect of actually taking the pill, even when compliance is imperfect. In this case, random assignment (being told to take the pill) serves as the instrument. It affects sleep quality only through its impact on whether people take the pill. Using this instrument, we can isolate the effect among those who comply, that is, those who take the pill when assigned to do so but would not take it otherwise. This gives us the **local average treatment effect (LATE)**, in other words the causal effect for compliers.

Instrumental variables are versatile tools. In practice, it is often not trivial to find a valid instrument. However, as mentioned, in the business world, there are several useful common situations, for example, when you can't randomize the treatment but can randomize incentives (encouragement or nudges).

Note that other quasi-experimental methods exist. However, these are the most popular. Moreover, the most important aspect of this book is to build a causal mindset, a way to understand and question causal claims. This set of examples illustrated the idea of a quasi-experimental method and revealed that our tools allow us to question these methods, in particular when we question the counterfactual and identify potential confounders.

To conclude this chapter, we will now turn to our usual case studies to apply the content of this chapter.

Case studies

Let's now apply the quasi-experimental methods we've just learned to our three running case studies. Your task is to suggest which approach seems most promising for finding a causal answer, based on the tools we've discussed.

More importantly, what would be the biggest challenge in applying that method? Remember, every quasi-experimental design hinges on a critical identifying assumption. As you read through the potential strategies for each case, focus on identifying that core assumption and thinking about how you might defend or discard it.

Case study 1: Cold showers and the immune system

To examine the effects of cold showers on health, quasi-experimental methods may not be the most effective approach. The challenge is finding a clean, real-world scenario. You would need to find a situation where some people are essentially forced to take cold showers while a similar group is not—for example, maybe a boiler failure in a large apartment building that lasts for months. But even then, this would not be very convincing, as maybe people shower less, heat water in the kitchen to bath, and so on.

Here is another idea. Cold exposure became popular in Switzerland in recent years. Often, near lakes in winter, you would find a sauna or showers, and changing rooms to facilitate swimming in cold water. These facilities might increase the probability of doing cold exposure while not forcing you to do so. Why not use the distance to such facilities as an instrumental variable? The closer you live to such facilities, the easier it is to go swimming in winter. The distance to a lake might be confounded with socio-economic status, as the rent near lakes might be more expensive, but you could restrict the study to people living near lakes, which would reduce this issue.

That being said, I see two potential weaknesses here. First, setting up such facilities near lakes is definitely not random. It might be built where there are potentially more people interested (reverse causation). Second, I suspect that it would be difficult to obtain granular data in order to set up such a study. Data availability is often a limiting factor in research.

Bottom line, I honestly didn't find any convincing setup or existing examples doing this, which is also a useful lesson. It is not always possible to find a good setup to use such methods. Hence, to measure the impact of cold showers on health, it seems that the best we had was a randomized experiment using objective physiological metrics, as discussed at the end of the last chapter.

Case study 2: Marketing campaign for energy sobriety

This case study focuses on the marketing strategy in Switzerland to reduce energy consumption. A potential way to evaluate the effectiveness of such a marketing strategy would be to use a difference-in-differences approach. Policymakers could gradually implement the campaign and compare its impact between similar regions where the policy is in place and where it has not yet been introduced. However, there are several challenges to this approach. First, people might travel between areas and be exposed to the campaign while working elsewhere. Second, strictly limiting the campaign geographically can be difficult, especially with online campaigns, at least at a sub-national level. Lastly, given the urgency of the situation, it could be hard to justify a gradual rollout of the campaign where some regions are not exposed to the campaign, as policymakers would likely aim to maximize its impact as quickly as possible.

Hence, another control group could be found in neighboring countries. However, this would obviously not be trivial to do either. While comparing entire countries might not be a valid comparison due to varied potential confounders, focusing on specific regions geographically might be better. You would need to obtain granular, weekly, or monthly energy consumption data from a neighboring country (such as Germany or France) for small geographical units. Then, you might be able to find a valid potential control group, comparing a region in Switzerland with a region in France, for example. It is essential to ensure that the control regions do not simultaneously implement their own energy-saving policies, which would be confounded with the effect of interest.

Case study 3: Air pollution and respiratory disease

This case study was discussed during the chapter, showing how difference in differences could be used to measure the effect of air pollution on respiratory disease.

Summary

We now conclude the third part of this book, which explored the main tools for inferring causality. Quasi-experimental methods proved to be particularly effective in addressing the limitations of randomized experiments. In particular, we can use such methods when randomized experiments are impossible or unethical. In addition, these methods usually allow us to study aggregated and longer-term effects, which are often more relevant than the short-term impacts measured in randomized experiments.

That being said, as we have seen in this chapter and throughout the book, we cannot prove causality with a simple test. Here, causal claims rely on the identifying assumption. If these assumptions are valid, it indicates that we may have identified a causal effect. To challenge this assumption, you must use the tools you have acquired, among other things, the concepts of parallel worlds and confounders, which correspond to steps 1 and 2 of the *Causal Mindset Framework*.

While we have seen how to measure a causal effect, we will now turn to an important yet potentially surprising last part: Even a perfectly measured causal effect can be misleading due to the choice of metrics and our interpretation.

Part 4

How to Deal with Imperfection

In the fourth and final part of the book, we will continue our ascent toward enlightenment, where we will learn that even when causality is correctly measured, we might be misled by the results due to blind spots. We will explore how perfectly sound causal analyses can mislead if they focus on the wrong question or metric, a problem known as the Type III error: answering the wrong question correctly. You will discover how the choice of measurement can make or break your conclusions, and how Goodhart's Law, *When a measure becomes a target, it ceases to be a good measure,* applies to real-world decision-making. Finally, we will embrace the imperfection inherent in all knowledge and see how developing a causal mindset is not about eliminating uncertainty but learning to navigate it wisely. By the end of this part, you will be able to identify limitations based on the choice of metrics and choose the right ones to answer your questions reliably, allowing you to deal with imperfection and fully exploit the strength of causal inference.

This part of the book includes the following chapters:

8

Correct Answer, Wrong Question: The Importance of Choosing the Right Metrics

After all the things we have studied together, let me take you through a last twist: causal estimates can be misleading. How is this possible? The main reason is that we tend to overinterpret the results or have blind spots due to the choice of metric used for the outcome. First, measuring a causal effect does not automatically explain how the effect happened. In a randomized experiment to test the effect of cold showers on health, you might be able to measure a causal effect on work-related sick leave. But is it due to an immune boost from temperature shock? Is it a mental resilience effect that lowers stress? Is it just a placebo effect? As we have seen previously, using reliable physiological metrics instead of a self-reported survey would address this issue. Second, we might not be able to measure the main outcome of interest, but just a proxy or another available and related variable. If we do not realize this, it often leads to heroic extrapolation. Hence, measuring the right outcome or at least assessing the limitations of using a proxy metric is central to reaching correct conclusions.

The preceding summarized problems reveal two fundamental ways a correct causal answer can be misleading: first, by misunderstanding the causal pathway (the "how"), and second, by measuring a proxy outcome that doesn't capture what truly matters and making a misleading outcome extrapolation. I like to frame this type of problem using terminology from statistics: "Type III error," which refers to answering the wrong question correctly.

Let me give you a quick example. A perfectly measured causal effect can backfire and even be misleading if the channel is misunderstood. For example, in the medical sector, the mortality rate of patients is sometimes used as an outcome to explore the quality of work. This is a natural conclusion—after all, doctors are here to save lives. Imagine that you want to measure the impact of training for doctors in a service on the mortality rate. After training, using proper causal inference methodology, you measure a strong reduction in the mortality of patients. But before popping the champagne, it might be worth diving into the "How?". Maybe the training was efficient. Or, if the doctors are aware that the mortality rate is now measured as a KPI, they might just start refusing risky patients or redirecting them to other services. If this is the case, while the causal effect shows a reduction in mortality, it is actually disastrous. The patients who require the most help, risky patients, might struggle to receive the aid needed.

This chapter will explore this subtle but central concept and will highlight how the choice of metric could be the issue and the solution.

In this chapter, we will cover the following topics:

- Answering the wrong question correctly
- Selecting the right metrics to measure what matters

Answering the wrong question correctly

Where are we and where are we going? These are potentially the most fundamental questions we can ask ourselves, ask regarding our company or professional projects, and even what politicians might wonder at the country level. In order to assess where we are and track our progress, an obvious solution is to measure, quantify, and, in other words, translate reality into numbers.

Numbers help to bring a sense of objectivity and facilitate comparison. If I say that we had a good financial year during a meeting, everyone wants to know the benefits and revenues in dollars. If I say that I have lost a lot of weight to my doctor, they need to know how much weight I have lost.

Numbers are powerful, but they can also mislead. In the very simple examples of benefits and weight loss, we can already see that two different types of information are required to make the information more complete. Absolute numbers are useful (e.g., benefits and weight loss), but the importance of the change depends on the relative terms with percentages. For example, losing 10 kg (22 lbs) is not the same for someone who weighs 100 kg (220 lbs) as for someone who weighs 50 kg (110lbs). Moreover, it might also be important to know the sex of the person, as well as their age. A seemingly simple question often can't be answered properly with a single metric. The scale, the reference points, and often other metrics are required to have a complete picture. When we turn to causal evaluation, it is central to also reflect precisely on the metrics.

If you don't pick the right metric, you might measure something causally that is actually not the right thing to measure. Unfortunately, as causal inference is getting more popular and coding is more accessible than ever, this is a common issue I have seen in my advisory practice. Some teams would identify some metrics, run a complex model with code, and get a result without any bugs. Running a model is simple nowadays. In particular, with generative AI, almost anyone can just write in natural language and get some code to do causal inference. The code generated would not have any bugs, but it doesn't mean that the model will be correct.

There are two main reasons why we might answer the wrong question correctly. Either we misunderstand the causal pathway, or we do misleading outcome extrapolations. Let us study both of these issues in detail.

Misunderstood pathway

The other day, I was discussing the effect of a research paper with a colleague. The paper is called *Durably reducing conspiracy beliefs through dialogues with AI*, and it reveals the fact that tools like ChatGPT can help to change people's beliefs about conspiracy theories (link: `https://www.science.org/doi/10.1126/science.adq1814`). This is a randomized experiment allowing us to pin down the causal effect. But now comes a second-order question: What explains this effect? This is called the channel. My colleague said it is because these tools are infinitely patient and don't have emotions, so they don't get angry. The research does not answer this question. Is that the reason? Is it because it is just very good at debating (which probably relates to emotional response)? Is it because it comes from a machine, and people have the impression that it is the truth? Understanding the channels or mechanisms for a causal effect is the natural follow-up and is central.

Having a rationale to explain the effect is necessary when you study causal inference. If you don't know what would explain the effect, it is hard to suggest a correct directed graph, identify potential confounders, and find a good proxy counterfactual. Actually, this is a weakness of the research I mentioned on cold showers and work-related sick leave. In the paper, they say: *"The mechanism or explanatory pathway of any therapeutic effects of cold exposure remains unclear."* (link: `https://pmc.ncbi.nlm.nih.gov/articles/PMC5025014/`).

Without knowing the channel through which the effect operates, the value of pinning down a causal effect may be greatly reduced or even misleading. Imagine that you provide high-tech training for a sales team using virtual reality and other exciting tools. You observe that the people (randomly selected) who followed the training outperformed others after the training. You might think that it is fantastic news.

However, what if the key channel driving this boost of productivity was not the content itself but the novelty effect? The employees felt valued by the company and their management, which motivated them. If this were the main driver, it is possible that redoing a similar type of training in a year or so would have no effect, as the novelty effect is not there anymore.

When pinning down a causal effect, understanding the channel helps to build confidence in the result and better forecast the impact of future actions. Disentangling the placebo effect, the Hawthorne effect, which refers to the potential change in behavior of a subject due to the fact that they are observed, and other common issues is necessary.

To pin down the issue, we are going to use causal graphs and will see two main ways to identify the problem with graphs. Then, by changing the metrics, we will be able to solve many issues.

We can use our toolbox to analyze and potentially solve this phenomenon. By making the directed graph more detailed, we can find the different potential pathways connecting the cause and the effect. In particular, decomposing the causal pathway from the cause to the outcome often reveals that we oversimplified the pathways and missed indirect effects or intermediate steps. These additional nodes on the causal pathway are called mediators, as we saw in *Chapter 1*. Let me get back to the example of high-tech training for a sales team.

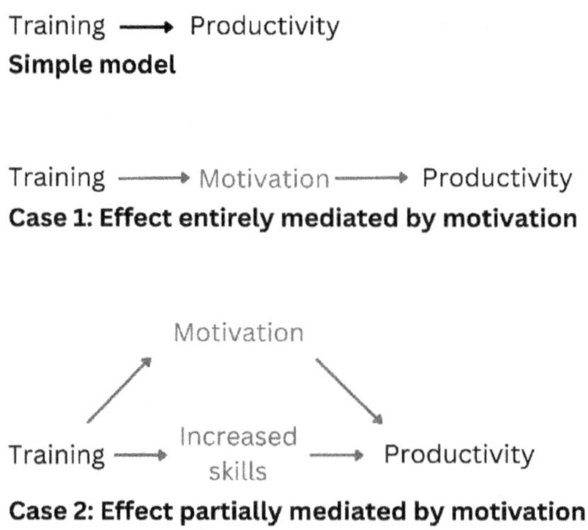

Figure 8.1: Directed graphs representing the issue of a misunderstood pathway when we do not observe the mediators, represented in gray

As we can see in the preceding figure, the simple model on top does not allow us to understand the "How?". The two other cases illustrate how mediators explain the causal effect and the two different possibilities. Either the mediator entirely explains the effect (case 1) or it is only one part of the story (case 2).

How do you know which situation you are in? Case 1, case 2, or case 3 (where the effect is entirely explained by increased skills)?

Another classic example would be the effect of education on earnings later in life. First, let's assume that we managed to measure a causal effect of one additional year of education on salary later in life. If you are interested in research measuring this causally, see, for example, the seminal paper using instrumental variables: *Does Compulsory School Attendance Affect Schooling and Earnings?* (link: `https://www.jstor.org/stable/2937954`).

However, there might be different pathways: Is it a story of knowledge, networking, or both? The classroom experience might arm students with specialized skills that directly boost their productivity. Simultaneously, networking, the relationships formed with peers, professors, and alumni, may foster valuable connections that lead to better job opportunities. Without unpacking these distinct mechanisms, we're unsure why exactly more schooling pays off, and risk designing policies or individual decisions that miss the true channels driving higher earnings.

There are several ways to answer this question, and as usual with causal inference, it is a good mix between technical knowledge and domain knowledge. The very first thing is to draw graphs, and drawing directed graphs relies primarily on domain knowledge. It is your expertise that will allow you to ask the right questions and suggest potential pathways. We will focus on this aspect in this chapter: how to use domain knowledge to select the right metric and answer the right questions.

In addition, there are two other more advanced technical sets of tools that could be used to support this type of analysis. These tools go beyond the content of this book. However, let me just share the intuition and resources to look further if you are interested. The first is called mediation analysis, and the idea is to study how the effect of interest is related to the different mediators in a data-driven way. To go further, you can refer to the following seminal papers in psychology: *The moderator–mediator variable distinction in social psychological research: Conceptual, strategic, and statistical considerations* (link: `https://pubmed.ncbi.nlm.nih.gov/3806354/`) and *Mediation in experimental and nonexperimental studies: new procedures and recommendations* (link: `https://pubmed.ncbi.nlm.nih.gov/12530702/`). Mediation analysis exploits statistical tools to explore how the relationship between your cause and outcomes varies when taking into account the different potential mediators.

The other related technical tool is called causal discovery. The starting point of any causal inference enquiry is the directed graph. However, this relies on domain knowledge and might be subjective in some situations. Hence, to complement this approach, the field of causal discovery aims to uncover causal relationships among multiple variables in a data-driven way, by analyzing statistical associations instead of relying on predefined causal structures. All the resources on causal discovery that I know are highly technical. The most accessible and reliable sources are the paper by Martin Huber *An Introduction to Causal Discovery* (link: `https://arxiv.org/abs/2407.08602`), and the book by Aleksander Molak, *Causal Inference and Discovery in Python*.

While these data-driven methods are powerful, I've witnessed in my own practice that they work best as a complement to, not a replacement for, domain knowledge. Hence, this is why I focus on a less technical approach in this book. Data-driven methods such as mediation analysis or causal discovery can spot statistical patterns, but they can't replace expert insight. You might have a polished dataset on training outcomes, yet fail to capture subtle but critical factors that only domain knowledge can help to identify, as no algorithm will spot what isn't measured. The best results emerge when you combine solid data science with your own on-the-ground understanding.

Before exploring the potential solutions, let me now turn to the second type of problem that could arise.

Misleading outcome extrapolation

Often, the causal effect is well measured, but we only observe a proxy metric. Without realizing it, we don't directly observe the variable of interest. Moreover, as the analysis is causal (high internal validity), we tend to be overconfident with this *small* extrapolation to the metric of interest. This is different from the problem of external validity, as discussed in *Chapter 4*.

External validity is about extrapolating—for example, in time or space—but using the same variable. Let me illustrate with a recent paper, which found that when a consumer or stakeholder becomes aware that a company uses generative AI for tasks such as advertising creation or product design, their trust in that company is reduced (see *The transparency dilemma: How AI disclosure erodes trust*—link: `https://www.sciencedirect.com/science/article/pii/S0749597825000172`). While the paper relies on 13 experiments and is arguably causal, I questioned its external validity: "Can we extrapolate this effect into the future? In other words, is it only a short-term effect before the world fully adapts to GenAI?" The concern is with the same variable, trust, but the comparison is between its impact today and its potential impact once GenAI becomes more integrated.

In this section, I want to discuss something different: extrapolation to a different outcome. In the example of the impact of GenAI on trust, we might be tempted to say that as GenAI erodes trust, it might also negatively affect sales or growth. While this sounds logical, we do not observe this, and it is important to note that there is some uncertainty.

Let me illustrate with a second example. While I was doing research on environmental impact, I looked into the metrics regulations and the type of metrics the companies had to report in varied **Environmental, Social, and Governance (ESG)** frameworks. Most of the time these metrics suffered from this issue of misleading outcome extrapolation. For example, one key metric in the automobile industry was the percentage of waste recycled. At first glance, this sounds great: recycling reduces environmental impact! But it's not that straightforward. If you only look at the percentage of waste recycled and assume it automatically benefits the environment, you are making the mistake of misleading outcome extrapolation.

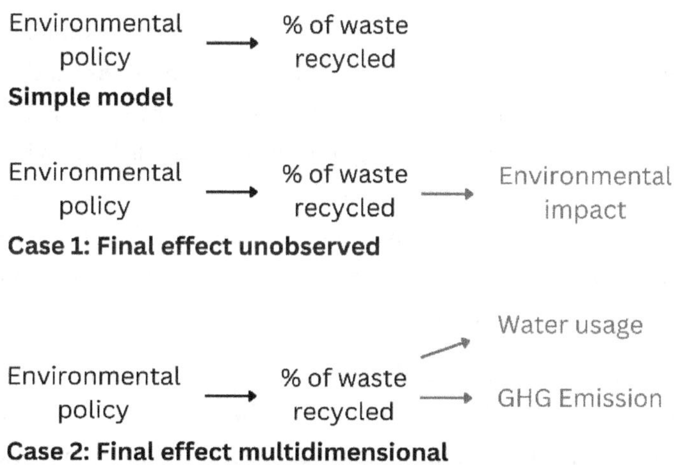

Figure 8.2: Directed graphs illustrating misleading outcome extrapolation, where the measured outcome (% of waste recycled) is observed, but in Case 1 the true outcome of interest lies further downstream and is unobserved, and in Case 2 the true outcome is multidimensional and unobserved (e.g., water usage and GHG emissions) and only partially captured by the observed metric

Recycling plastic, glass, or leather, locally or in different countries, has a different environmental impact. Here, there is an oversimplifying assumption that a higher percentage of waste recycled necessarily leads to lower environmental impact. The issue of misleading outcome extrapolation might take two different forms.

Let's start with case 1, as illustrated in *Figure 8.2*. In this scenario, while the percentage of waste recycled is increasing, it is possible that the overall quantity of waste is also increasing. With simple numbers, this could be something like this: Initially, the company has a total amount of waste of 100 tons, and 10 tons are recycled, which leads to a 10% recycling rate. Now imagine that the company has 150 tons of waste and recycles 20 tons. The company is recycling 13.3% of the waste; this metric has increased. However, the total amount of waste increased from 100 to 150 tons, and the total amount of unrecycled waste increased as well, from 90 to 130 tons.

The second situation is illustrated in *Figure 8.2* with case 2. Another potential issue arises when the metric is too narrow. Having a causal mindset requires using systems thinking. It helps to understand the big picture and should include direct and indirect pathways as well as direct, indirect, and multi-dimensional outcomes. In the case of waste management, recycling might require water as well as energy. Hence, it is also important to consider all the potential dimensions as we might do better on one, while doing worse on another one. This arises again if we oversimplify the pathway, leaving blind spots when making misleading outcome extrapolations.

Similarly, this is also the case with GenAI implementation. A fintech company made headlines in 2025 after saying that they were going AI-first and reducing their marketing teams. They decided to re-hire people a few months later, as it was apparently not such a good idea. The issue of misleading outcome extrapolation might have played a role here. When using GenAI for marketing, you might increase productivity and reduce costs. But as we saw at the beginning of this section, it could also erode trust or reduce quality. Hence, both of these issues might affect client satisfaction and customer loyalty. These metrics are probably what matter the most, as they could be the main drivers of long-term growth.

This is exactly why these concepts relate to the problem of answering the wrong question correctly. It is essential to focus on what truly matters and ask whether you are actually measuring it, or just a proxy that may not affect your real outcome of interest in the way you expect.

We've seen how the wrong metric can mislead us. Let's now turn to how metrics themselves can become part of the solution.

Selecting the right metrics to measure what matters

Both issues—misleading outcome extrapolation and misunderstood pathways—arise from missing information. In the first case, information about the causal pathway (the mediator) is ignored. In the second, information further down the chain of effects is not observed (another outcome of interest). One way to resolve this issue is to carefully select the metric you want to capture in order to see what would have been unseen, and hence avoid answering the wrong question correctly.

Let us first focus on the issue of misunderstood pathways. If we take the example of the high-tech training for the sales team, measuring motivation, satisfaction, and skills acquired in a follow-up survey could help to mitigate the issue by revealing whether the effect was entirely explained by motivation or also by reinforced skills.

Let's now tackle the second type of problem: misleading outcome extrapolations. This issue is also often addressed by changing the outcome, how we measure the outcome, or studying the impact on multiple outcomes.

In the case of a new environmental policy measured by the percentage of waste recycled, we saw how it can be misleading to assume that a higher recycling rate automatically translates into a lower environmental impact. Once the issue is identified using the causal graph, you can adjust the metric and instead focus on a more comprehensive measure, such as the carbon footprint of waste management (e.g., CO_2-equivalent emissions). This metric captures the entire environmental impact from waste management, taking into account the cost of recycling, of increasing landfill, of transporting waste, and so on. Hence, now the metric is aligned with the goal and prevents blind spots. In addition, you could also use several metrics to avoid having a narrow focus—for example, including metrics about water usage or the pollution of freshwater.

Hence, the natural solution is to use directed graphs to make potential blind spots visible. Sometimes, of course, the information simply isn't available. Still, mapping out the assumptions in a clear graph is essential, as it helps us accept uncertainty and anticipate potential issues.

For example, let's say you are doing online marketing, and your goal is to increase the **Click Through Rate (CTR)**, which is the proportion of people clicking a link when they see it. This is particularly important on YouTube to optimize thumbnails. However, over-focusing on such metrics might be shortsighted and lead to destructive strategies. You can use a clickbait strategy where you create a high incentive to click the link, but when watching the video, viewers are disappointed about the content and never come back to your channel. Hence, even if you are capable of measuring the causal effect of your choice of thumbnail on CTR, it might be a risky metric to focus on exclusively, leading to bad incentives and long-term negative effects. In this situation, it might be difficult to track the longer-term effect of a particular thumbnail on brand loyalty or even company growth. Recent research explores the possibility of inferring long-term effects from short-term metrics to address this type of limitation (see *The Surrogate Index: Combining Short-Term Proxies to Estimate Long-Term Treatment Effects More Rapidly and Precisely*, link: `https://academic.oup.com/restud/advance-article/doi/10.1093/restud/rdaf087/8268796`).

There is a final central aspect to discuss when we choose our metrics and KPIs: behavioral biases.

Assessing and addressing behavioral bias

When people know that a specific **Key Performance Indicator (KPI)** is tracked, they might forget the real goal and over-focus on the metrics, leading to manipulations or counterproductive actions. There is even a name for the phenomenon where a metric, once turned into a target, ceases to be a reliable measure: Goodhart's law.

Many anecdotal stories well illustrate this issue. Imagine that a government wants to reduce the number of venomous cobras in a region and hence offers a reward for every dead cobra. While this might encourage the killing of snakes first, it might also create incentives to breed cobras to claim more rewards. Similarly, some weapons buyback programs, where people are rewarded for turning in their guns to reduce availability and violence, have faced criticism. They can create incentives to hand in old or broken weapons for a reward, without meaningfully reducing the number of functional guns in circulation or the level of violence.

Another example I mentioned earlier is the problem in the medical sector of using mortality rate as a KPI. As the mortality rate becomes the KPI, people might do anything to improve this metric—in this case, reducing it even if it means refusing to treat risky patients who need treatment.

Similarly, in the case of sustainability, some companies might voluntarily, or not, harm the environment by over-focusing on the metric. The percentage of waste recycled was such an example.

In academia, a common issue is that people are evaluated on the number of papers published and citations. Indirectly, the quality is assessed by looking at the journal where the papers are published. But overall, this issue is at the core of the adage *publish or perish*, which drives people to always publish more, at a faster pace, often manipulating data or using other tricks to get published.

While the previous section essentially explains how to reduce this risk by using more objective or complete metrics, there is an additional concept that can help you anticipate such behavioral bias. This idea was initially shared by my good friend Prof. Charles Ayoubi, and uses terminology from the legal sector. Charles told me that the issue in academia is that there is an obligation of results rather than an obligation of means.

First, let's define these terms. Obligation of results means that you have a duty to deliver a good result and are evaluated on the result, while obligation of means focuses on the means and procedure in place and not the results. Many of these issues arise due to a failure to choose the right metrics aligned with the type of obligations. In academia, we should have an obligation of means. We should be evaluated on the quality of the procedure, of the effort, and not only on the outcomes.

If you are evaluated on the outcomes, it might push people to manipulate the data—for example, simply for the sake of publishing more. Extreme cases such as the amyloid beta scandal, where a Stanford professor manipulated data, arguably for the sake of publishing more and getting citations in the research into Alzheimer's, leading to countless follow-up research papers, vast amounts of money being spent, and probably a lot of time being lost.

This is exactly the same issue with several examples I shared before: the cobra story, the gun buyback programs, or the mortality rate in the medical sector. In the medical sector, mortality is an outcome, and hence this implies an obligation of results, eventually driving counterproductive incentives. On the other hand, if you focus on the standard of care and the quality of the procedure, you will focus on the means and not the results.

One way to address this is to focus on metrics robust to manipulation. In the example of cold showers, eventually focusing on objective physiological measures instead of self-reported behavior would be safer.

However, it is not always possible to obtain such metrics robust to manipulations. Hence, when designing KPIs, it's important to align how people are measured with what they can genuinely control. If your goal is to encourage ethical behavior, thorough procedures, and sustained quality over time, you might favor an obligation of means, rewarding well-executed efforts rather than strict outcomes. Conversely, if your main concern is achieving quantifiable results (such as hitting specific profit or subscription targets) and you have confidence that individuals truly can influence those results, an obligation of results could be more appropriate.

Here are some guidelines:

- **Assess the level of control**: If external factors (e.g., market fluctuations, patient health conditions, or peer-review randomness in academia) heavily influence outcomes, it's often fairer and more constructive to focus on the means. On the other hand, if team members have a high degree of control over the end goal, then an obligation of results can ensure accountability.
- **Anticipate behavioral responses**: An obligation of results can push individuals to manipulate data or cut corners if the final outcome is overly emphasized, as illustrated with the academia and medical sector examples. An obligation of means can reduce such perverse incentives, but if it's too loose or vague, it risks complacency; people might not strive to achieve the final outcome since they're measured mostly on procedure, not performance. Hence, often a mix of metrics is necessary to obtain a balanced perspective. For example, you can evaluate a researcher on proper methodology (means) and publication record

(results). A balanced set of KPIs ensures nobody feels compelled to sacrifice ethics or quality for the sake of short-term wins. Of course, this is not trivial, and multiplying the metrics for evaluation might also make the analysis more complex.

- **Set clear, appropriately matched KPIs:** When setting KPIs, align them with the type of obligation required. For obligation-of-means roles, use process-focused metrics such as checklists, quality reviews, or peer assessments to evaluate adherence to best practices. For obligation-of-results roles, rely on outcome-focused metrics such as sales targets, subscriber growth, or measurable performance indicators, ensuring they accurately reflect what individuals can influence.

In essence, deciding between an obligation of means and an obligation of results boils down to who can influence what and how. When someone has direct control over an outcome, linking success to that outcome can be highly motivating. But if complex factors escape a person's control, emphasizing procedural quality (means) may be both fairer and more effective. By thoughtfully aligning KPIs with either means or results (or both), you create an environment that promotes true performance without driving destructive or manipulative behavior.

Case studies

Let's now apply the concepts from this chapter to our three running case studies. For each scenario, we will look beyond the initial causal finding and ask the critical follow-up questions: Do we understand how the effect is working? Are we truly measuring what matters? This exercise will show how choosing the right metrics and understanding the causal pathway is essential for moving from a correct answer to a truly wise one.

Case study 1: Cold Showers and the immune system

We have already covered this situation in this chapter, in the *Assessing and addressing behavioral bias* section. In our ongoing cold-shower scenario, you might find a causal effect on reduced work-related sick leave. Measuring physiological measures (e.g., blood samples, immune markers) would allow us to ensure that the cold showers truly affect the outcome of interest: health.

Case study 2: Marketing campaign for energy sobriety

Imagine now that we have finally managed to find a valid quasi-experimental setting to causally measure the impact of the energy sobriety campaign in Switzerland. The result reveals a causal reduction in energy consumption. Now, we can ask: "What is the causal pathway?"

Suggested answer: It would be interesting to study whether people followed the habits presented in the campaign (e.g., boiling water with the lid on a pan or reducing the thermostat and wearing a sweater) or used other methods to save energy. In addition, studying the driving force with surveys or interviews would be interesting. Indeed, if the driving force was just newly acquired knowledge, this might allow for a longer-term habit change and long-lasting effect, which would also benefit the environment. But if the effect was driven mainly by fear of shortages or price increases, the impact may quickly fade once the crisis passes or energy supplies stabilize. Tracking these additional metrics would help to confirm the causal pathway and evaluate the potential longer-term impact.

Case study 3: Air pollution and respiratory disease

The paper on air pollution reduction policies around the Olympic Games in China helps isolate the causal impact of pollution on health. However, let's challenge this result in light of this chapter's concepts:

Suggested answer: The policy reduced air pollution and mortality from respiratory diseases. But we can ask if air pollution reduction is the only causal pathway explaining this effect. What if, in Beijing, where the policies are implemented and where the Olympic Games take place, there is an increase in the availability and quality of medical services to accommodate foreigners and the increased number of people?

If this were the case, we would probably observe a reduction in mortality from any type of disease. However, the author reveals that there is no impact on other varied outcomes: non-cardiovascular disease or injury, for example. This is a good illustration of how using varied metrics allows us to shed light on the potential causal pathway.

In addition, what if people are suddenly inspired by the Olympic Games and start doing more sports, which eventually improves health and reduces mortality from cardiovascular or respiratory diseases? The author actually mentions this limitation: *"The [pollution reduction policy] may have affected people's preferences between indoor and outdoor activities. The consequences of such behavioral changes on mortality merit further investigation."*. Again, tracking other metrics might help to evaluate the importance of this alternative pathway—for example, fitness club occupation, sports club subscription, public activities (bike-share rides), sales of sports gear, or data from a running platform such as Strava.

Summary

Here is a final word on this topic. By now, we've seen that identifying a causal effect might not solve all problems and could still be misleading. To avoid running blindly in the wrong direction, we must dig into how it happens, pick the right metrics, and align those metrics with the type of obligation—means or results—that makes sense. When we zoom in on potential mediators, we gain clarity about the true drivers behind a result. By checking for misleading outcome extrapolations, we ensure our conclusions hold up outside the neat confines of a single measurement. Ultimately, a careful mix of domain knowledge, robust methodology, and adequate KPI design makes the difference between a good decision and a costly misstep.

You may have now reached the stage that most participants in my classes, workshops, or keynotes eventually get to: Can we really be sure that something is causal? Let me address this point in the next and final chapter.

Join our community on Discord

Join our community's Discord space for discussions with the authors and other readers: `https://discord.gg/G4YmB6DQvf`

9

Embracing Uncertainty

While teaching The Causal Mindset in a top engineering school to master's students the other day, someone raised their hand and said: *How is it possible that such an important concept is so widely misunderstood?* I suspect that now you are also surprised by this situation. Hopefully, this book shows that we can change this status quo.

When you started the book, you may have been unsure about causal inference and even probably thought that "causality" felt like an obscure technical concept, not really important. Then, hopefully, we quickly realized how central causality is to decision-making in our daily lives, from personal decisions to geopolitical strategies. We discovered that while AI is powerful and at the core of many discussions and investments today, it is not the right toolbox for answering causal questions, as it focusses on prediction using patterns and correlations.

Of course, there was a catch: the fundamental problem of causal inference. We fell into this valley of despair, probably thinking, "This is why we don't hear so much about causal inference." But slowly and surely, we climbed the slope of enlightenment. First, by dissecting the challenges: the choice of **the control group or proxy counterfactual**, confounders, and reverse causation.

Second, we kept climbing towards more clarity by studying the solutions. We saw that building a causal mindset and applying the framework presented in this book to make a qualitative assessment of a causal claim on the spot can go a long way. Then, we turned to more advanced solutions: randomized experiments. Still questioning everything, we witnessed the strengths and limitations of such experiments. But luckily, we discovered another toolbox, quasi-experimental methods, which allow us to complement elegantly randomized experiments on our quest to answer causal questions.

Finally, we saw how the choice of metrics is a subtle task with profound impact. Metrics can either shed light on the causal pathway and effect or obscure and mislead you. This last part revealed the importance of using causal inference with systems thinking, allowing you to identify the varied causes, multiple pathways, and eventually multidimensional outcomes.

Over the years, while teaching the causal mindset to thousands of people, I witnessed two different reactions at this stage. The first is that you are more confident with your decisions. The second is that you are more uncertain than ever. Let me address this situation, in case you are experiencing it. The thing is, without opening this book or the box of causal inference, many people don't even realize that they are making mistakes. They live in a simpler world, which might be very pleasant. However, this oversimplification is costly, as we have seen throughout the book. But now, you can't go back to that simple world.

Understanding causality and questioning it should not prevent you from making any decisions. You should feel empowered by the new skill. Instead of living in a misleading, simplified world, you can now accept the complexity and know how to navigate it with your causal mindset. There are arguably no simple answers to complex questions. I often say the same thing about statistics and causal inference: It is the humbling and much-needed art of dealing with uncertainty.

Mastering causality and developing a causal mindset does not mean having all the answers; it means learning how to deal with uncertainty. We've begun climbing the slope of enlightenment and covered a lot of ground, but this kind of knowledge takes time to sink in. We must get used to seeing the world through this new lens.

Essentially, causality is not black or white. Causal evidence can be viewed rather on a spectrum, from anecdotal pieces of evidence to perfectly measured causal effects. Anecdotal pieces of evidence bring some information, yet very weak information: I took a pill to help me sleep yesterday, and it gave me the best night's sleep I've had in a long time. That's a single observation with no comparison group, so we can't tell whether the pill, coincidence, or simply a calmer evening caused the effect. While we never truly have the perfect causal estimation with the parallel worlds, we can get very close to this situation. In the case of sleeping pills, randomized experiments that compare sleep quality using objective physiological indicators between a treated group and a placebo group provide solid evidence, assuming the study does not suffer from the issues we discussed in *Chapter 6*. The randomization ensures that, on average, the two groups are similar in every important respect (e.g., stress, diet, genetics). The use of a placebo allows researchers to isolate the effect of the pill itself from psychological expectations—people's belief that they are being treated can already improve their sleep. Using objective physiological measures rather than self-reports prevents key biases such as desirability bias or memory distortion, providing a more reliable estimate of the true effect. While this is eventually one robust piece of evidence, we usually pile up more evidence from different sources with different strengths and weaknesses, eventually gaining more confidence about a specific topic and effect.

Some people argued to me that spreading this knowledge has some risks. It might encourage skepticism and rejection of the scientific evidence. However, I disagree with this statement. It even sounds like the church in the Middle Ages, willing to prevent people from reading to keep asserting its power by possessing exclusive access to knowledge. Education is key to empowering people against manipulation. In addition, as I have shown in the entire book, scientific evidence remains the best that we have. Sure, it is imperfect, and now you have acquired tools with this book to challenge findings and, for example, the interpretation of scientific results by the general public, in the news, or on social media. Far from being destructive, this kind of scrutiny is healthy. It acknowledges uncertainty without preventing decisions, and it makes us more careful, which is exactly what the complexity of our world demands.

What now? Let me first point to further resources. For those who want to study the more technical aspects of causal inference, I invite you to check Matheus Facure's book *Causal Inference in Python* or *Causal Inference The Mixtape* by Scott Cunningham.

Most importantly, do not just put this book aside and forget about it. This book offers tools, case studies, and a structured method, yet real mastery comes from using these ideas, from practicing them in your daily life, whether you're deciding on a marketing campaign, launching a policy, or just rethinking your morning routine. True mastery will come with practice.

And so, as we part ways, I invite you to keep questioning everything with an unrelenting curiosity. Continue refining your approach, whether it's a small tweak in daily decisions or an entire strategy in public policy. The world needs more people who look beyond surface correlations, simple monocausal explanations, and short-term fixes.

If there is one thing I really want you to do after closing this book, it is to take note every time you hear, read, or say therefore or because. At that moment, a causal assumption is made and could be questioned with the content of this book. Simply noticing how often we make or hear such claims without reliable evidence should be enough to convince you to apply these concepts in your daily life.

Together, armed with knowledge and our causal mindset, we can face the challenges of our times, not just as individuals, but as a collective force. In a society polluted by fake news, misleading messages on social media, or in politics, it is a much-needed skill to acquire and share. Hence, do not keep this knowledge for yourself. Share it. You'll see that simply asking simple questions ("What is the counterfactual? Is there something else? Could it be the reverse, or can we extrapolate?") will expose others to this way of thinking, eventually improve their decisions, and contribute to changing the status quo.

Get this book's PDF version and more

Scan the QR code (or go to `packtpub.com/unlock`). Search for this book by name, confirm the edition, and then follow the steps on the page.

UNLOCK NOW

Note: Keep your invoice handy. Purchases made directly from Packt don't require an invoice.

10

Unlock Your Exclusive Benefits

Your copy of this book includes the following exclusive benefits:

DRM-Free PDF Version

Download DRM-free PDF and ePub copies of this book.

7-Day Packt Library Access

Get 7-day unlimited access to 8,000+ books and videos. No credit card required.

Available for first-time Packt+ trial users only.

Next-Gen Reader Access

Read this book on Packt Reader with progress sync, dark mode and note-taking.

Follow the guide below to unlock them. The process takes only a few minutes and needs to be completed once.

Unlock this book's free benefits in 3 easy steps

Step 1

Keep your purchase invoice ready for *Step 3*. If you have a physical copy, scan it using your phone and save it as a PDF, JPG, or PNG.

For more help on finding your invoice, visit `https://www.packtpub.com/en-us/unlock?step=1`.

> **Note:** If you bought this book directly from Packt, no invoice is required. After *Step 2*, you can access your exclusive content right away.

Step 2

Scan the QR code or go to `packtpub.com/unlock`.

On the page that opens (similar to *Figure 10.1* on desktop), search for this book by name and select the correct edition.

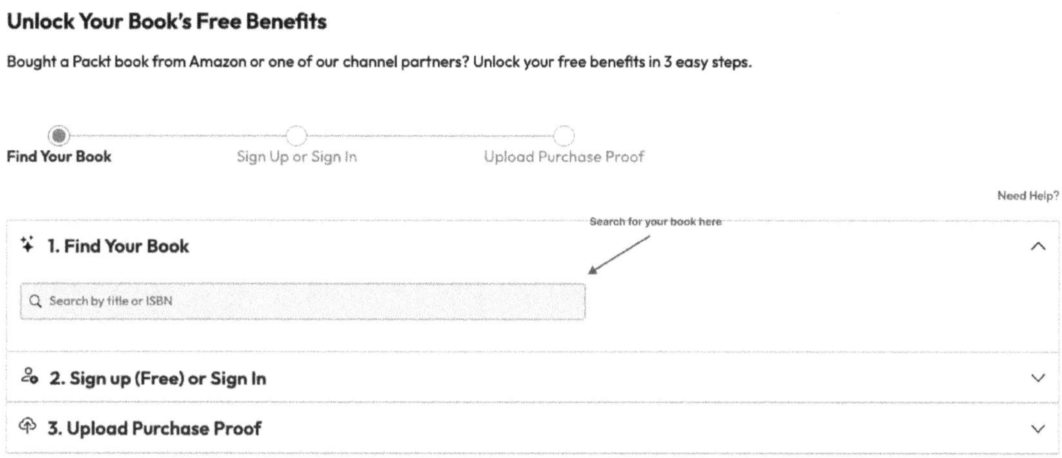

Figure 10.1: Packt unlock landing page on desktop

Step 3

After selecting your book, sign in to your Packt account or create one for free. Then upload your invoice (PDF, PNG, or JPG, up to 10 MB). Follow the on-screen instructions to finish the process.

Need Help

If you get stuck and need help, visit `https://www.packtpub.com/unlock-benefits/help` for a detailed FAQ on how to find your invoices and more. This QR code will take you to the help page.

Note: If you are still facing issues, reach out to `customercare@packt.com`.

Other Books You May Enjoy

If you enjoyed this book, you may be interested in these other books by Packt:

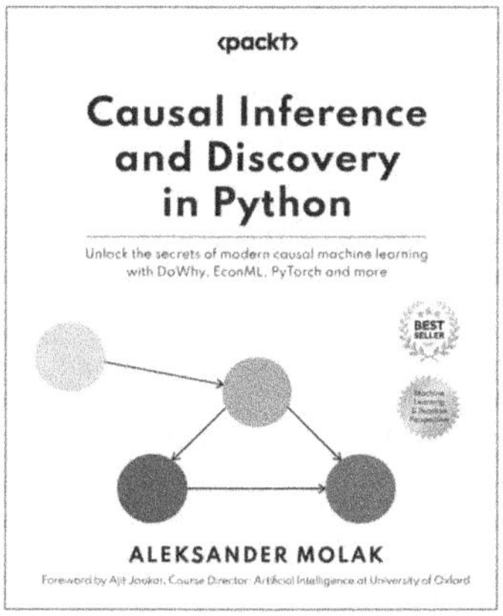

Causal Inference and Discovery in Python

Aleksander Molak

ISBN: 978-1-80461-298-9

- Master the fundamental concepts of causal inference
- Decipher the mysteries of structural causal models
- Unleash the power of the 4-step causal inference process in Python
- Explore advanced uplift modeling techniques
- Unlock the secrets of modern causal discovery using Python
- Use causal inference for social impact and community benefit

Causal Inference in R

Subhajit Das

ISBN: 978-1-83763-902-1

- Get a solid understanding of the fundamental concepts and applications of causal inference
- Utilize R to construct and interpret causal models
- Apply techniques for robust causal analysis in real-world data
- Implement advanced causal inference methods, such as instrumental variables and propensity score matching
- Develop the ability to apply graphical models for causal analysis
- Identify and address common challenges and pitfalls in controlled experiments for effective causal analysis
- Become proficient in the practical application of doubly robust estimation using R

Packt is searching for authors like you

If you're interested in becoming an author for Packt, please visit `authors.packt.com` and apply today. We have worked with thousands of developers and tech professionals, just like you, to help them share their insight with the global tech community. You can make a general application, apply for a specific hot topic that we are recruiting an author for, or submit your own idea.

Share your thoughts

Now you've finished *The Causal Mindset Handbook*, we'd love to hear your thoughts! Scan the QR code below to go straight to the Amazon review page for this book and share your feedback or leave a review on the site that you purchased it from.

`https://packt.link/r/1806117851`

Your review is important to us and the tech community and will help us make sure we're delivering excellent quality content.

Index

D

E

F

G

H

I

K

L

M

N

O

P

Q

R

S

T

V

W

www.ingramcontent.com/pod-product-compliance
Lightning Source LLC
LaVergne TN
LVHW060035210926
848481LV00013B/3870
9781806117857